Organized
Networks of
Carbon Nanotubes

Organized
Networks of
Carbon Nanotubes

Edited by

K.R.V. Subramanian
Aravinda C.L. Rao
Raji George

CRC Press
Taylor & Francis Group
Boca Raton London New York

CRC Press is an imprint of the
Taylor & Francis Group, an **Informa** business

CRC Press
Taylor & Francis Group
6000 Broken Sound Parkway NW, Suite 300
Boca Raton, FL 33487-2742

First issued in paperback 2021

ISBN 13: 978-1-03-223977-4 (pbk)
ISBN 13: 978-0-367-27820-5 (hbk)

Visit the Taylor & Francis Web site at
www.taylorandfrancis.com

and the CRC Press Web site at
www.crcpress.com

Contents

Preface

Carbon nanotubes (CNTs) are cylindrical shells made, in concept, by rolling graphene sheets into a seamless cylinder. CNTs exist as either single-walled nanotubes (SWNTs) or multiwalled nanotubes (MWNTs). Iijima's report in 1991 helped create awareness of CNTs to the scientific community and triggered a deluge of interest in CNTs. A variety of methods for the production of aligned arrays in catalytic chemical vapor deposition has been demonstrated. Horizontally well-aligned CNT arrays on suitable substrates are highly desired for various SWNT device applications, such as field effect transistors, sensors, light emitters, and logic circuits.

Regularly bent CNTs are proposed for applications including mechanical nanospring devices, high-resolution atomic force microscopy tips, and nanocircuit interconnections in device. Self-assembly can form some novel structures. Ensembles of CNTs form networks that can be transparent and at the same time conductive, showing great potential for replacing the currently used indium tin oxide (ITO), whose future availability is at risk due to the impoverishment of the world's sources of indium. CNT networks offer not only an alternative to ITO but also advantages such as flexibility, usability with plastic substrates, and capability to avoid contamination from ITO of other device materials. In this book, we expand and elucidate some novel and state-of-the-art techniques used for alignment of CNTs and their composites, to form organized networks. The initial chapter dwells upon properties and characterization of CNTs, while subsequent chapters deal with the novel and interesting alignment methodologies and their applications.

We, the editors, Subramanian, George, and Aravinda Rao, draw upon our extensive research experience and works in the field of organized CNT networks to infuse this book with the required bent and perspectives. We would like to express our sincere thanks to our management, namely Ramaiah Institute of Technology, Bangalore, India (Shri M.R. Jayaram, Chairman, GEF; Shri M.R. Seetharam, Vice Chairman, GEF and Director, Ramaiah Institute of Technology; Dr. N.V.R. Naidu, Principal, Ramaiah Institute of Technology) as well as Reliance Industries Ltd., Vadodara, India, for supporting this endeavor. We also thank the distinguished authors who have contributed their mite. We finally thank our family members for their support.

Professor K.R.V. Subramanian
Professor Aravinda C.L. Rao
Professor Raji George

Author Biography

Professor K.R.V. Subramanian is Professor in the Mechanical Engineering Department and also works as Co-Research Coordinator, Ramaiah Institute of Technology, Bangalore, India. His research interests are solar technology, nanotechnology, carbon nanotubes, fuel cells, energy storage, and sensors. He earned his PhD from Cambridge University in 2006 with specialization in nanotechnology. His bachelors and masters degree in materials engineering were from NIT, Trichy and IISc, Bangalore, respectively. He has over 20 years of academic and industrial work experience. He has published over 100 journal and conference papers. He has also edited 2 books. He is a Marquis Who's Who of the World and IBC Top 100 professionals. He has been co-principal investigator for many government funded research projects.

Dr. Aravinda C.L. Rao is working as a General Manager at Product Application & Research Center, Reliance Industries Ltd., Vadodara. After completing his doctoral studies at Bangalore University in 2001, he pursued his post-doctoral work at IISc, Bangalore, as an Alexander Humboldt fellow at Karlsruhe Institute of Technology, Technical University of Munich, Germany, and University of California at Riverside. His research interests encompass nanotechnology, graphene-based advanced materials, polymer composites, biosensors, thin films, and coatings for electronics. He has rich and varied academic as well as industrial R&D experience of over 18 years in multinational companies. He has authored more than 25 papers in reputed journals, a book chapter, and several conference papers. He has handled large to medium scale R&D projects successfully and has established and managed large research labs.

Dr. Raji George is a Professor and Head in the Mechanical Engineering Department of Ramaiah Institute of Technology (RIT), Bangalore. His areas of interests include subjects such as nanotechnology, mechanics of materials, and design and finite element methods. He earned his PhD from Visvesvaraya Technological University (VTU) in 2008. He has more than 33 years experience in academia and teaching. He was awarded Scientific Award of Excellence of 2011 by American Biographical Institute. He was conferred the Best Engineering Teacher Gold Medal Award for the year 2009, instituted by Sir M. Visvesvaraya Memorial Foundation, Bangalore. He has more than 40 journal and conference papers to his credit. He has 1 patent granted in his name and applied for 1 patent. He is the Principal Investigator for a major research project for Boeing Inc.

Contributors

B.V. Raghu Vamsi Krishna
Department of Mechanical
  Engineering
GITAM University
Bengaluru, India

Janardan Sannapaneni
Department of Chemistry
GITAM University
Bengaluru, India

Sannapaneni Janardan
Department of Chemistry, Gitam
  School of Science
GITAM University
Bengaluru, India

S. Krishna Prasad
Centre for Nano and Soft Matter
  Sciences
Bengaluru, India

D.S. Shankar Rao
Centre for Nano and Soft Matter
  Sciences
Bengaluru, India

Marlin Baral
Centre for Nano and Soft Matter
  Sciences
Bengaluru, India

G.V. Varshini
Centre for Nano and Soft Matter
  Sciences
Bengaluru, India

Aishwarya V. Menon
Center for Nano Science and
  Engineering
Indian Institute of Science
Bangalore, India

Tanyaradzwa S. Muzata
Department of Materials
  Engineering
Indian Institute of Science
Bangalore, India

Suryasarathi Bose
Department of Materials
  Engineering
Indian Institute of Science
Bangalore, India

Sushant Sharma
CSIR-National Physical Laboratory
New Delhi, India
Academy of Scientific & Innovative
  Research (AcSIR)
Ghaziabad, Uttar Pradesh, India

Abhishek Arya
CSIR-National Physical Laboratory
New Delhi, India
Academy of Scientific & Innovative
  Research (AcSIR)
Ghaziabad, Uttar Pradesh, India

Sanjay R. Dhakate
CSIR-National Physical Laboratory
New Delhi, India
Academy of Scientific & Innovative
  Research (AcSIR)
Ghaziabad, Uttar Pradesh, India

Bhanu Pratap Singh
CSIR-National Physical Laboratory
New Delhi, India
Academy of Scientific & Innovative
  Research (AcSIR)
Ghaziabad, Uttar Pradesh, India

Brijesh Kumar Mishra
Computational Chemistry Unit
International Institute of
  Information Technology
Bangalore, India

Balakrishnan Ashok
Centre for Complex Systems & Soft
  Matter Physics
International Institute of
  Information Technology
Bangalore, India

Dr. K.R.V. Subramanian
Department of Mechanical
  Engineering
Ramaiah Institute of Technology
Bangalore, India

A. Deepak
Department of Mechanical
  Engineering
GITAM University
Bangalore, India

Dr. Aravinda C.L. Rao
Product Application & Research
  Center, Reliance Industries Ltd.
Vadodara, India

Dr. T. Nageswara Rao
Department of Mechanical
  Engineering
GITAM University
Bangalore, India

Dr. Raji George
Department of Mechanical
  Engineering
Ramaiah Institute of Technology
Bangalore, India

1

Properties of Carbon Nanotubes

B.V. Raghu Vamshi Krishna

Department of Mechanical Engineering
GITAM School of Science GITAM University
Bengaluru, India

Sannapaneni Janardan

Department of Chemistry
GITAM School of Science GITAM University
Bengaluru, India

1.1 Introduction

Carbon nanotubes (CNTs) are rolled up graphene sheets, generally of two types, namely single-walled CNTs (SWCNTs) and multiwalled CNTs (MWCNTs). There are variations in both SWCNTs and MWCNTs. They vary by length, purity, and practicality. CNTs are long, thin carbon wires just a nanometer or so in diameter; however, they can be up to several thousands of times in length. They possess exciting mechanical, electrical, and optical properties that would make them ideal nanoscale materials. Rolling of CNTs is described with circumferential vector denoted by **Ch** that is expressed in terms of primitive vectors of graphene sheet [1].

$$\mathbf{Ch} = n\mathbf{a1} + m\mathbf{a2}$$

where **a1** and **a2** are primitive vectors in 2D hexagonal lattice in Figure 1.1. Generally, CNTs are classified into three groups according to the values of 'm' and 'n'. For armchair nanotubes ($n,m = n$ and chiral angle = 30°), carbon–carbon bonds are perpendicular to the axis of tube, for zig-zag nanotubes (either n or m are zero or chiral angle = 0°), carbon–carbon bonds are parallel to the axis of tube as shown in Figure 1.2, and for chiral tubes (n,m and chiral angle 0° < θ < 30°) [1].

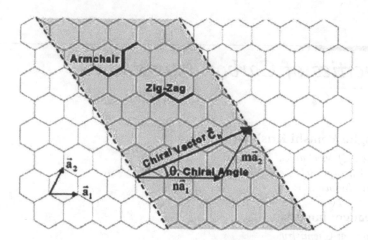

FIGURE 1.1
Unit vectors **a1** and **a2** and armchair, zig-zag, and chiral configuration [1].

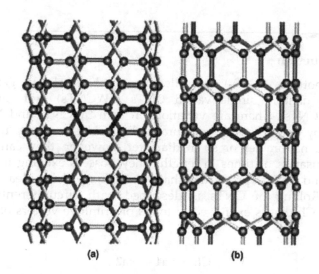

FIGURE 1.2
Armchair and zig-zag CNTs [1].

1.2 Properties of CNTs

1.2.1 Electronic Properties

Electronic properties of CNTs can be altered by changing its structure (n-m vector). If the vector is a multiple of 3 then the CNTs are metallic (conductor) and if the *n-m* vector is not a multiple of 3 then the CNTs are semiconductors [2]. Two main properties, Fermi energy (E_F) and energy gap (E_g), are calculated to investigate the properties of chiral CNT. Energy gap is the difference between the orbitals, that is, from the lowest unoccupied orbital (LUMO) to highest occupied orbital (HOMO), and this energy gap is caused by the curvature of tube wall. Electrical conductivity of CNTs is determined by energy gap; the smaller the gap, the stronger the conductivity. E_F and E_g are defined in equations 1 and 2.

$$E_F = (E_{LUMO} + E_{HOMO})/2 \tag{1}$$

$$E_g = E_{LUMO} - E_{HOMO} \tag{2}$$

If the vector (*n*, *m*) is not a multiple of 3 then the electronic density of states of SWCNTs shows the band gap near fermi level and produces a semiconductor. And if the vector (*n,m*) is a multiple of 3, both valance band and conduction band of SWCNTs overlap each other, producing a conductor.

From Figure 1.3, it is observed that chiral nanotube of vector (8, 4) and (10, 5) is direct bandgap semiconductor and (4, 2) is indirect band gap semiconductor. CNTs of vector (6,3) and (12,6) are conductors with energy gap of 0.33 and 0.1 eV. Metallic nanotubes are highly conductive as they can carry billions of amperes of current due to few defects and their resistance is also very low [3].

1.2.2 Mechanical Properties

Many experiments have been conducted to investigate the properties of CNTs. Because of sp^2 hybridization of C–C bonds, it is expected to have

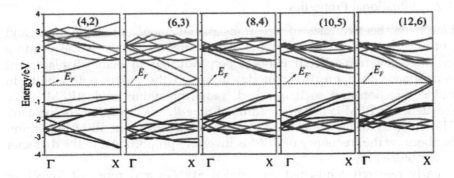

FIGURE 1.3
Energy band structures of chiral SWCNTs [3].

very high Young's modulus (in terms of TPa), breaking strength, stiffness, and tensile strength. These properties depend on chirality and size. The Young's modulus of CNTs for different diameters ranges from 1.28 to 1.8 TPa, when compared to steel (0.21 TPa) it is about 10 times stronger [4].

Treacy et al. determined the mechanical properties of SWCNTs by thermal vibration by TEM method. The Young's modulus $E = 1.77 \sim 1.37 / +2.38$ TPa for $d_o = 5.6 \sim 24.8$ nm, $d_i = 1.0 \sim 6.6$ nm, $t = 3.3 \sim 18.2$ nm [4,5]. Lu has determined the mechanical properties of SWCNTs by Empirical force-constant model method. The properties are $E = 0.97 \sim 1.11$ TPa, $v = 0.269 \sim 0.280$, $G = 0.436 \sim 0.541$ TPa, for $d = 0.68 \sim 27.0$ nm [4,6]. Properties of CNTs determined by different methods are given in Table 1.1

In MWCNTs, the stress transferability between layers is weak and the maximum load is taken by the outermost layer. So at the outermost layer of MWCNTs, the failure will start by breaking the bonds between carbon–carbon atoms [4].

Graphene and CNTs (armchair, zig-zag, and chiral) are modeled in analysis system (ANSYS), and Young's modulus was determined analyti-cally. As the diameter of the chiral nanotube increases, there is a slight increase in Young's modulus, and as the diameter of the zig-zag nanotube increases, there is an increase in Young's modulus approximately from 0.99 to 1.05 TPa. So the mechanical properties change as the diameter and chirality of CNTs change [14].

Because of the bending of CNTs, there is a significant change in atomic structure from sp^2 to sp^3 hybridization. Because of this, the electrical conductance of SWCNTs is greatly reduced. For armchair (n, n) CNTs, there is no change in bandgap and their electrical conductivity remains same. But for zig-zag $(n, 0)$ and chiral nanotubes, the bandgap increases due to increase in tension and because of this the conductivity of CNTs may change from conducting to semiconducting due to deformation shown in Figure 1.4. [15].

1.2.3 Vibrational Properties

Graphene sheet consists of two translational phonons with in-plane and out-of-plane atomic displacements. As the graphene sheet is rolled into a cylinder, the translational phonon perpendicular to plane displacement corresponds to a breathing mode, in which the atoms will vibrate in radial direction and so it is called 'radial breathing mode (RBM)'. The frequency of RBM is linear to radius of tube R, $\omega(R) = C/R$, C is constant. The diameter of the nanotube can be calculated from the RBM wavenum-ber because the frequency of RBM is inversely proportional to the diameter of nanotube [16].

Early research conducted on tangled SWNTs and ropes of SWNT or MWNTs is measured by Raman Spectra. Raman spectra exhibit three bands: the RBM band at ~ 200 cm^{-1}, that is, the vibration at which the

TABLE 1.1

Mechanical properties of CNTs, which are determined by different methods

CNT	Method	Properties
SWCNTs	Analytical Asymptotic Homogenization Model [7]	$E = 1.44$ TPa, Shear modulus = 0.27 TPa, Tube thickness = 1.29 Å
SWCNTs	Numerical–structural Mechanics-based 3D Finite Element Model (FE) approach [7]	$E = 0.97{\sim}1.05$ TPa, Shear modulus = 0.14~0.47 TPa, Tube thickness = 6.8 Å
Double-walled CNTs	Numerical–structural Mechanics-based FE approach [7]	$E = 1.32{\sim}1.39$ TPa, Shear modulus = 0.37~0.62 TPa, Tube thickness = 6.8 Å
MWCNTs	Numerical–structural Mechanics-based FE approach [7]	$E = 1.39{\sim}1.58$ TPa, Shear modulus = 0.44~0.47 TPa, Tube thickness = 6.8 Å
SWCNTs	Modified Morse Potential [8]	$E \approx 1.13$ TPa, $\sigma_{TS} = 126.2$ GPa for armchair, $\sigma_{TS} = 94.5$ GPa for zig-zag
SWCNTs	Finite Element Analysis Beam Element Model [9]	$E = 1.005{\sim}1.04$ TPa $G = 0.155{\sim}0.455$ TPa For armchair $d = 0.5{\sim}2.8$ nm $E = 1.027{\sim}1.081$ TPa $G = 0.32{\sim}0.48$ TPa For zig-zag, $d = 0.5{\sim}2.8$ nm $E = 0.965{\sim}1.031$ TPa $G = 0.32{\sim}0.48$ TPa
MWCNTs	Atomic Force Microscopy [10]	$E = 350 \pm 110$ GPa $G = 1.4 \pm 0.3$ GPa $d_0 = 20{\sim}50$ nm, $L > 0.8$ μm
SWCNTs	Molecular Mechanics Model–Modified Morse Potential. Reactive Empirical Band Order Potential [11]	For zig-zag $\sigma_{TS} = 105.38$ GPa, $\varepsilon_f = 0.33$ For armchair $\sigma_{TS} = 134.01$ GPa, $\varepsilon_f = 0.43$
MWCNTs	Timoshenko Beam Model [12]	For MWCNTs $E = 126{\sim}937$ GPa, $G = 33{\sim}785$ MPa, $d_0 = 30{\sim}60.9$ nm, $d_i = 7.9{\sim}12.4$ nm, $L = 1.80{\sim}18.4$ mm
SWCNTs	Molecular Mechanics [13]	For armchair $E = 0.963{\sim}1.025$ TPa SWCNTs $v = 0.15{\sim}0.29$ For zig-zag and chiral, $d = 0.4{\sim}1.8$ nm, $L = 1.8130{\sim}2.8986$ nm, $E = 1$ TPa for graphene

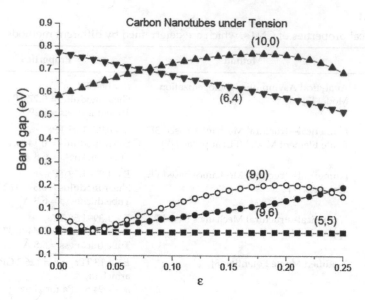

FIGURE 1.4
The variation of bandgap as the strain increases for CNTs.

diameter of nanotube expands and contracts, the G band at ~1600 cm^{-1}, that is, the in-plane vibration of graphite, and weak disorder band (D band) at ~1350 cm^{-1}. Figure 1.5 shows the Raman Spectra for ropes of SWCNTs at different laser excitation energies and it can be observed that there is a

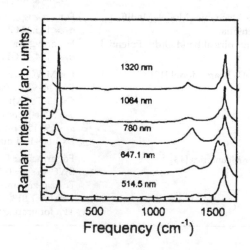

FIGURE 1.5
Raman spectra for ropes of SWNTs at five different laser excitation energies.

sudden increase in peak intensity. For samples of nanotube usually of different chirality and radius, the Raman spectra gives information about diameter of nanotubes and their chirality. So Raman spectra is an important tool for structural characterization [16].

References

1. Paradise M; Goswami T. Carbon nanotubes – production and industrial applications. *Materials and Design* 28 (2007) 1477–1489.
2. Yanagi K. *Differentiation of Carbon Nanotubes with Different Chirality*. Department of Physics, Tokyo Metropolitan University, Hachioji, Tokyo, Japan (2014).
3. Liu J; Lu J; Lin X; Tang Y; Liu Y; Wang T; Zhu HJ. The electronic properties of chiral carbon nanotube. *Computational Materials Science* 129 (2017) 290–294.
4. Hu Z; Lu X. *Mechanical Properties of Carbon Nanotubes and Graphene*. Elsevier (2014).
5. Treacy MMJ; Ebbesen TW; Gibson JM. Exceptionally high Young's modulus observed for individual carbon nanotubes. *Nature* 678–680 (20 June 1996) 381.
6. Lu JP. Elastic properties of carbon nanotubes and nanoropes. *Physical Review Letters* 79 (7) (18 August 1997) 1297.
7. Kalamkarov AL; Georgiades AV; Rokkam SK; Veedu VP; Ghasemi-Nejhad MN. Analytical and numerical techniques to predict carbon nanotubes properties. *International Journal of Solids and Structures* 43 (2006) 6832–6854.
8. Xiao JR; Gama BA; Gillespie JW, Jr. An analytical molecular structural mechanics model for the mechanical properties of carbon nanotubes. *International Journal of Solids and Structures* 42 (2005) 3075–3092.
9. Kirtania S; Chakraborty D. Finite element based characterization of carbon nanotubes. *Journal of Reinforced Plastics and Composites* 26 (15) (2007) 1557–1570.
10. Guhados G; Wan WK; Sun XL; Hutter JL. Simultaneous measurement of Young's and shear moduli of multiwalled carbon nanotubes using atomic force microscopy. *Journal of Applied Physics* 101 (2007) 033514.
11. Duan WH; Wang Q; Liew KM; He XQ. Molecular mechanics modeling of carbon nanotube fracture. *Carbon* 45 (2007) 1769–1776.
12. Wei XL; Liu Y; Chen Q; Wang MS; Peng LM. The very low shear modulus of multi-walled carbon nanotubes determined simultaneously with the axial Young's modulus via in situ experiments. *Advanced Functional Materials* 18 (10) (2008) 1555–1562.
13. Ávila AF; Lacerda GSR. Molecular mechanics applied to single-walled carbon nanotubes. *Materials Research* 11 (3) (2008) 325–333.
14. Lu X; Hu Z. Mechanical property evaluation of single-walled carbon nanotubes by finite element modelling. *Composites: Part B* 43 (2012) 1902–1913.
15. Liu B; Jiang H; Johnson HT; Huang Y. The influence of mechanical deformation on the electrical properties of single wall carbon nanotubes. *Journal of the Mechanics and Physics of Solids* 52 (2004) 1–26.
16. Popov VN. Carbon nanotubes: properties and application. *Materials Science and Engineering R* 43 (2004) 61–102.

2

Characterization of Carbon Nanotubes

Sannapaneni Janardan

Department of Chemistry, Gitam School of Science
GITAM University
Bengaluru, India

B.V. Raghu Vamshi Krishna

Department of Mechanical Engineering
GITAM School of Science GITAM University
Bengaluru, India

2.1 Introduction

The innovation of CNTs occurred conceptually in early 1950s and 1970s. However, their significance was highlighted in 1990s by Iijima. Initial investigations suggested that SWCNTs have rolled up graphene as a single sheet, whereas the MWCNTs have similar structure as SWCNTs but with more concentric tubes in multiple numbers [1–3]. Among the three types of CNTs (zig-zag, armchair, and chiral), chiral form is having significant optical, mechanical, and electrical properties [4, 5].

The CNTs usually possess ballistic electron transport throughout the nanotubes due to their 1D structure at room temperature. In addition, they possess high thermal conductivity (>3000 W/m K) than diamond and basal plane of graphite [6]. Moreover, because of their high Young's modulus, they possess high mechanical properties over other steel materials [7]. CNTs have shown significant applications in the fields of physics and mechanics [8], hence they are involved in manufacturing of devices such as nanomanipulation, hydrogen storage, and nanoporous membranes [9]. There are several unexplored fields that have potential application in aerospace, automobile, composite reinforcements, catalyst supports, probes, sensors, and nanoreactors [10]. In addition, characterization techniques are highly significant in measuring the defects occurred during purification process [11]. In this chapter we address the significant characterization techniques that are used to determine the complex CNTs – both quantitative and qualitative analysis. To identify the large number of problems that have occurred with respect to controlled production and to elucidate the morphology and structure of CNTs, in Table 2.1 various kinds of characterization techniques and their usages are described.

TABLE 2.1

Different analytical techniques used for the characterization of CNTs [12]

Characterization Technique	Used for Studying
Microscopy and Diffraction Techniques	
AFM	Morphological analysis of internal structure (diameter, number of layers, and distance between them)
TEM	Morphological analysis of internal structure (diameter, number of layers, and distance between them)
SEM	Morphological analysis of internal structure (diameter, number of layers, and distance between them)
ND	Morphological analysis of bulk samples
XRD	Morphological analysis of bulk samples
Spectroscopic Techniques	
RS	Purity and presence of by-products, diameter distribution (n, m), and chirality
IR and FTIR	Purity, functionalization by attaching functional groups to the sidewalls
UV–Vis and NIR	Dispersion efficiency, diameter and length distribution, and purity
FS	Size, dispersion efficiency (n, m), and chirality
XPS and EDS	Elemental composition and functionalization (covalent and noncovalent)
Thermal Techniques	
TGA	Purity and presence of by-products, quality control of synthesis and manufacture process
Separation Techniques	
SEC	Purification and separation by size (length)
Capillary electrophoresis	Purification, separation by size (length, diameter, and cross-section)
Field flow fractionation	Functionalization by size (length)
Ultracentrifugation	Separation by chirality, electronic type, length, and enantiometric identity

Among the above characterization techniques, the significant techniques are listed below.

2.2 Characterization Techniques for CNTs

2.2.1 X-Ray Photoelectron Spectroscopy

It is one of the significant techniques used for the identification of chemical structure of the CNTs. Generally, the SWCNTs that are free from the dopants consist of three peaks: (i) sp^3 carbon peak at 285 eV, (ii) sp^2 carbon peak at 284.3 eV, and (iii) sp^3 carboxyl group (C–O) peak at about 288.5 eV [13]. In case of MWCNTs, the sp^3 carbon energy position is about 284.6 eV, and 0.3 eV negative shift occurred between SWCNTs and MWCNTs clearly reveals the information about the weaker C–C bond binding energy and the interlayer spacing. The shifts in the binding energy always vary with the diameter of the nanotubes in MWCNTs, and for 30 nm diameter aligned MWCNTs, binding energy is about 50 meV between the sidewalls and tips.

2.2.2 Electron Microscopy

To observe the size, shape, and structure of the CNTs, electron microscopy (SEM, TEM) is one of the essential techniques. These techniques usually work based on the Lamberts law and analyze the CNTs' cross-section intensity. TEM is used to determine the linear adsorption coefficient and inner and outer diameter of MWCNTs [14]. MWCNTs intershell spacing, which is generally varying with diameter of CNT (0.34–0.39 nm), was identified by TEM. STM is a technique used to experimentally find out the energy band models and density of states, and it will give a clear picture about the electron densities with its structure [15]. To identify any kind of CNT structure, always coat the CNT on the conducting substrate and perceive the structure at low temperatures (4.2 K) [16]. Chirality and its influence on electronic properties of CNTs are generally determined by STM. In addition, it is also used to determine the properties of intramolecular junction and for measuring the electronic contacts, doping, defects and symmetry. Atomic vacancies (lateral size 0.5–0.8 Å, height 0.1 nm) of the metallic and semiconducting CNTs are also identified effectively by Computer simulation coupled STM [17].

2.2.3 X-Ray Diffraction

The statistical characteristics of CNTs such as diameter, distribution of chirality, number of layers, internal spacing, impurities, and structural strain were efficiently identified by XRD from Table 2.2, and the

TABLE 2.2

Characterization of pure, OH, and COOH functionalized MWCNTs

Type of CNTs	2θ (°)	Characteristic Peaks	Interlayer Spacing (d_{002}) (m)	Crystalline Size (D) (m)	[Ref]
Pure MWCNTs	25.6 40.0 59.0 71.0	002	3.47×10^{-10}	5.2×10^{-9}	[20,21]
(-OH) MWCNTs	26.2	002	3.47×10^{-10}	3.3×10^{-9}	[21]
(-COOH) MWCNTs	26.1	002	3.44×10^{-10}	3.5×10^{-9}	[21]

pattern of CNTs resembles the pattern of graphite, in which MWCNTs show peak at (002) clearly identified from Figure 2.1, which is also present in the XRD pattern of graphite, and the intensity of the peak always depends upon the alignment of the CNTs, that is, the reduction of the intensity of the peak (002) represents that the CNTs are properly

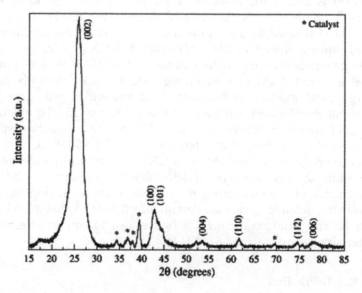

FIGURE 2.1
XRD spectra of MWCNTs.

Source: Evaluating the characteristics of multiwall carbon nanotubes, J. H. Lehman et al., Carbon, 49(8) (2011).

aligned [18]. However, this technique is not useful in differentiating the CNTs and graphite [19].

2.2.4 Raman Spectroscopy

This technique efficiently determines the tube alignment, defects, and purity as well as distinction of the MWCNTs relative to other carbon allotropes. In addition, it significantly describes the structure of SWCNTs. It is a nondestructive technique used to detect the purity of the CNTs and determine the properties of CNTs through its vibration modes resulting from the scattered photons released from the sample that is bombarded with intense light beam. It generally produces two kinds of bands: (i) graphitic band (G band) observed at around 1500–1600 cm^{-1}, which originates from the graphene sheets' vibration planes. (ii) Defect band (D band) observed at around 1300–1350 cm^{-1}, which originates from the sidewall of CNTs with defects. In addition, the ratio of defect band and graphitic band (D/G) will clearly give idea about the quality of the nanotubes, that is, if the D/G ratio is small then the CNTs are having minimum defects in their structure [22–26]. However, aggregation of CNTs will influence the D, G band signals, so in order to overcome the interferences of the aggregation of CNTs, treating with surfactants such as Triton X-100 is useful [27]. The diameter and chiral angle integers, n and m determine the structure and size of the CNTs. Based on different n,m values, the structure of the CNTs also varies from zig-zag, chiral, and armchair [28].

The characteristic peaks of SWCNTs in the Raman spectrum have lots of features, which include a frequency peak at <200 cm^{-1}, which is absent in graphite. The peaks that are obtained at 1340 cm^{-1} represent the disorder in the graphite material, which is known as D band having chirality due to double resonance conditions [29]. Moreover, in individual SWCNTs, D band generally splits into two bands and the separation always depends upon the incident light [30]. The multiple peaks obtained in 1550–1600 cm^{-1} range generally correspond to G band. For graphite, a single peak is obtained at 1582 cm^{-1}, which represents the graphitization of the sample and is classified into G+ and G− bands. G+ band represents the atomic displacement in the direction of axis of CNT, but in G− band, displacement is done in the circumferential direction and it will vary for metallic and semiconducting systems. The line obtained at 2600 cm^{-1} is the overtone of the D band, which is also called as G^{I} band and represents the long-range order in the sample. G band is the significant characteristic feature in distinguishing both SWCNTs and MWCNTs, which is decomposed into two bands in SWCNTs and it is asymmetric in MWCNTs [31].

2.2.4.1 Raman Spectra of CNT

From Figure 2.2, it is clearly possible to distinguish the MWCNTs (S0) and their functionalized forms (S1, S2, S3) through chemical oxidation

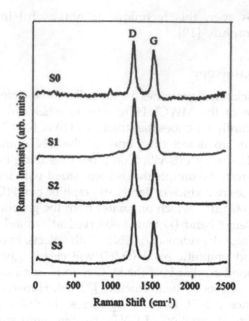

FIGURE 2.2
Raman spectra of pristine MWCNTs (S0) and their oxidized forms (S1, S2, S3).

Source: Characterization of multi-walled carbon nanotubes functionalized by a mixture of HNO₃/H₂SO₄, Le Thi Mai Hoa, Diamond and Related Materials, 89 (2018) 43.

by using HNO_3 and H_2SO_4 at various reaction parameters, which revealed that the peaks of D band for pristine MWCNTs, that is, S0, and oxidized MWCNTs, that are, S1, S2, S3 are at 1328, 1331, 1330, 1332 cm^{-1}, respectively. The G band for pristine MWCNTs (S0) and their oxidized forms (S1, S2, S3) was obtained at 1580, 1582, 1581, 1583 cm^{-1}, respectively in Table 2.3. The significant higher frequency shifts of oxidized samples over normal pristine MWCNTs were mainly due to

TABLE 2.3

Raman spectral data of pristine MWCNTs and their oxidized forms [24].

Sample	HNO₃ and H₂SO₄ Ratio	Treatment Time (h)	D Band Location (cm⁻¹)	G Band Location (cm⁻¹)	Id/Ig Ratio
S0	No treatment	No treatment	1328	1580	1.15
S1	1:3	6	1331	1582	1.20
S2	1:3	12	1330	1581	1.29
S3	1:3	18	1332	1583	1.40

doping through oxidation. The basic peaks of MWCNTs are retained even after oxidation, which indicates the presence of graphite structure as such after oxidation. The structural defect that is measured by the ratio of D and G bands (Id/Ig) is determining the presence of larger proportion of sp^3 carbon, which in turn is proportional to the defects and degree of disorder in the MWCNTs. The percentage of defects is increased with increase in the treated time, which can be easily observed through Table 2.3.

2.2.5 Absorption Spectroscopy

2.2.5.1 FT-IR Spectrophotometer

This technique is used to determine the quantitative and qualitative analysis of samples. Moreover, it is highly significant in determining the functional groups attached to the CNTs and the intensity of the peak increases with increase in the OH functional groups on the CNTs [32].

2.2.5.2 FT-IR Spectrum of CNT

The FT-IR spectra of MWCNTs in Figure 2.3 has shown characteristic peaks at 1096 cm^{-1}, which represent C–O stretching frequency, and the peak resonates at 1632 and 1634 cm^{-1} represent C=C stretching frequency and peaks at 2926 and 2856 cm^{-1} represent CH$_2$ asymmetric and symmetric stretching and the absorptions at 3436 cm^{-1} represent OH stretching. There is a comparative view of FT-IR spectra for CNT, oxidized CNT, and HAP-CNT represented in Table 2.4 [33]. The peak present with the 3000–3600 cm^{-1} is associated with hydroxyl group due to CNT oxidation. The C=O group resonates at 1735cm^{-1} (carbonyl acid stretching frequency), C–O–H in-plane bending at 1435 cm^{-1}, C–O–H out of plane bending at 995 cm^{-1}, C–O stretching at 1232, 1139, and 1067 cm^{-1}, which have resemblances with Silverstein et al. [27]. The peak of C=C resonates at 1624 and 1497 cm^{-1} (vibrations of aromatic nucleus), 889 and 846 cm^{-1} (adjacent or aromatic C–H vibration), and 660 and 560 cm^{-1} bands represent polar clusters attached to the CNTs aromatic ring. The phosphate group resonates at 560, 601, and 961 cm^{-1}; the crystalline apatite arrangement was vibrated at 1037 and 1095 cm^{-1}. The carbonyl absorption was shifted to 1740 cm^{-1}, which represents the successful bonding of COOH with HAP cation, and oxidized CNT vibrations were completely merged with phosphate vibrations at 958, 599, and 567cm^{-1}.

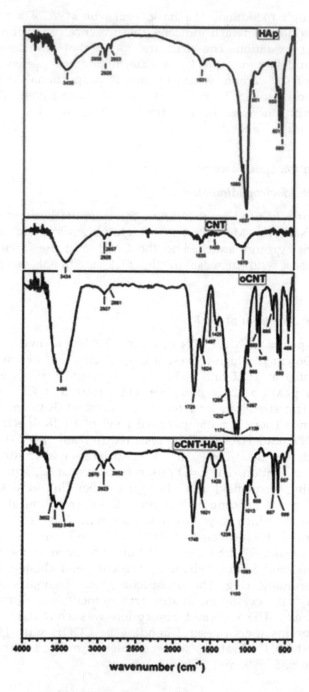

FIGURE 2.3
IR spectra of HAP, CNT, oCNT, oCNT-HAP.

Source: Oxidized-sulfonated multi-walled carbon nanotube/hydroxyapatite hybrid particles: Synthesis and characterization, Kaio A.B. et al., Journal of Solid State Chemistry, 279 (2019) 120924.

TABLE 2.4

Vibrations of CNT, oxidized CNT, HAP, and oCNT-HAP composite

	CNT	Oxidized CNT (oCNT)	HAP Functiona- lized CNT	oCNT-HAP Composite
-OH stretching	3434 cm^{-1}	3495 cm^{-1}	3435 cm^{-1}	3000–3600 cm^{-1}
CH$_2$/CH* asymmetric and symmetric stretching	2928 and 2857 cm^{-1}	*889 and 846 cm^{-1}	2926 and 2853 cm^{-1}	2923 and 2852 cm^{-1}
C=C stretching	1632 and 1634 cm^{-1}	1624 and 1497 cm^{-1}	1631 and 1430 cm^{-1}	1621 and 1420 cm^{-1}
C–O stretching	1096 cm^{-1}	1232, 1139, and 1067 cm^{-1}	1725 cm^{-1}	1740 cm^{-1}

* CH asymmetric stretching is observed only in oCNT.

2.2.5.3 UV-Vis NIR Spectroscopy

It is always necessary to quantify the SWCNTs in the dispersed form in solutions because all the applications of the SWCNTs depend upon the concentration of the CNTs, which in turn depends upon the activity of the CNTs. The concentration of the individually dispersed SWCNTs and the concentration of total dispersed SWCNTs and their complexity in the solution are measured by the absorbance and the resonance ratio of UV-Vis-NIR absorption spectrum. The wavelength range from 300to 600 nm is suitable for measuring the total concentration of the SWCNTs in the dispersed medium. In addition, the wavelength below 800 nm is suitable for estimating the individually dispersed SWCNTs through dilution experiments [34].

However, in case of MWCNTs, dispersion in aqueous surfactant solution is also measured by UV-Vis spectroscopy. Moreover, the maximum dispersion of MWCNTs is proportional to the absorbance in UV-Vis spectroscopy and it was determined by time-dependent sonication methods. In addition, with increase in the concentration of surfactants there is a gradual increase in the dispersion of MWCNTs and it requires less amount of sonication energy for dispersion and higher MWCNTs require more amount of total sonication energy for significant dispersion. The efficient dispersion will occur when the weight ratio of surfactant and MWCNTs is about 1.5–1. Moreover, the surfactants maintain the stability of MWCNTs for several months by restricting the accumulation of MWCNTs by absorbing on them and 1.4 wt% is the best ratio (surfactant to MWCNTs) where the MWCNTs are homogenously dispersed in aqueous solution [35].

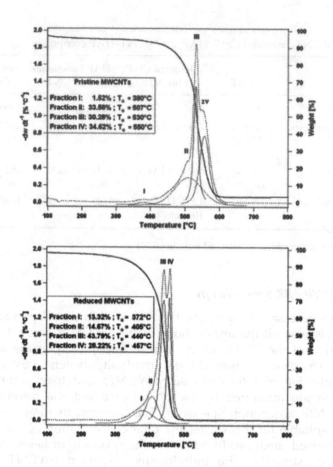

FIGURE 2.4
TGA curves of the pristine MWCNT (above) and reduced MWCNTs (below) [31].

Source: Characterization methods of carbon nanotubes: a review, T. Belin et al., Materials Science and Engineering: B, 119(2) (2005) 105.

2.2.6 Thermogravimetric Analysis

This is a sample destructive technique used to determine the quality of the CNTs (carbon metal ratio or SWCNTs/MWCNTs ratio) and their thermal stability based on the burning temperatures [36, 37]. However, no sample preparation is required for this technique. The weight loss curve of TGA generally consists of three major parameters of interest: (i) the initiation temperature, (ii) the oxidation temperature, and (iii) the residue mass. The initiation temperature describes the initial decomposition of the sample, the oxidation temperature describes the maximum amount of decomposition of the sample (representing thermal stability of the sample), and the residue mass describes the residual compound obtained after complete decomposition of the TGA process, which is mostly representing the catalyst used in the synthesis of CNTs and the decomposition temperature of the samples always depends upon the various factors such as impurities and defects found in the CNTs, number of walls, and catalyst used in the synthesis process [23, 24].

Among other carbon forms (graphite, fullerene, diamond, and soot), crystalline MWCNTs are more resistant to oxidation process. However, aromatic bonding is the key feature that significantly influences the thermal stability, which depends upon various factors such as defects in the tube, catalyst composition (used in the preparation of CNT), number of walls, amorphous carbon, and graphite particles. The width of the peak always helps in measuring the purity of the material; if the peak width is narrow then the material is highly pure. Moreover, with the less temperature there is significant liberation of the amorphous carbon in MWCNTs [38, 39]. The decomposition order will always follow in a sequence like amorphous carbon (200–300°C), SWCNTs (350–500°C), MWCNTs (400–650°C), and graphite particles. Scheibe et al. determined the decomposition curves of multiwalled CNTs, which result in oxidized and reduced forms. Oxidation temperature for the pristine tube is 300°C, which has resemblances to the pristine tube, and reduced form is at temperature 250°C due to the defects and functional groups incorporated on the tubes [40]. Differential TGA curve clearly shows four major fractions in which I represents amorphous carbon and the rest three fractions (II, III, and IV) represent CNT from Figure 2.4. Annealing process, modification in the nanotube, size, length, and diameter play a major role in influencing the oxidation temperature. However, the length, diameter, modification in the nanotube enhance the oxidation temperature [41, 42].

2.2.7 Neutron Diffraction

In order to investigate CNTs at large extent in terms of bond length and distortions in hexagonal network, one may require neutron diffraction [43]. This technique is highly significant in determining the structural features of the molecules such as armchair, zig-zag, and chiral nanotubes even with smallest diameter. Burien et al. has showed the C–C bond length of CNTs as 0.141 nm [44], Dore et al. determined two types of bond lengths, which are close to the graphite value and this distortion value measures the single and double bond character over the CNTs [45]. Moreover, the interlayer spacing of MWCNTs (0.345 nm) is greater than the graphite (0.335 nm). The size, number, and diameter of the nanotube bundles influence the neutron diffraction values, which are determined by comparing the theoretical values with practical values by Giannasi et al. [46] and they have classified the diffraction values into two Q regions: (i) for MWCNTs/bundle packing, it is $Q < 2°A^{-1}$, and (ii) for SWCNTs $Q > 2°A^{-1}$. In addition,

TABLE 2.5

Characteristics of Commonly Used Spectroscopic Techniques for Nanotube Characterization [12]

Technique	CNT Type/ Synthesis Method	Methodology Used for Sample	Wave Length Peak/ Range Studied	Note	Ref.
RS	MWCNTs	MWCNTs were dispersed in cyclohexane followed by sonication for 30 min	D band ($1342\ cm^{-1}$) and G band ($1580\ cm^{-1}$)	Excitation wavelength: 514.5 nm	[47]
RS	SWCNT and MWCNT	CNTs were dispersed in 5% (wt/vol) TritonX-100 solution followed by ultrasonication (10 min, 750 W)	D band around 1350 cm^{-1}, G band around 1593 cm^{-1},	Excitation wavelength: 532 nm	[48]
FT-IR	MWCNTs modified with 2-amino benzothiazole	MWCNTs were oxidized with a 3:1(vol:vol) mixture of concentrated H_2SO_4/HNO_3 prior to functionalization	$4000–400\ cm^{-1}$	FT-IR spectra were measured on KBr pellets	[49]
FT-IR	MWCNTs modified with ethylenediamine	MWCNTs were oxidized with HNO_3(c) and refluxed for 12 h with stirring prior to functionalization	$4000–400\ cm^{-1}$	FT-IR spectra were measured on KBr pellets	[50]
UV-Vis	MWCNTs	CNTs were dispersed by oxidation with a 3:1 mixture of H_2SO_4: HNO_3 and ultrasonication (2 min, 540 W)	253nm	The method was applied to study the influence of CNT concentration on the growth of fibroblasts	[51]
UV-Vis	SWCNT (HiPCo)	CNTs were dispersed in different 0.1 to 10% aqueous solutions of diverse surfactants	273 nm	Surfactants studied included SDS, CTAT, CTAB, NaDDBS, and SC	[52]
UV-Vis-IR	SWCNT	CNTs were dispersed in different solutions of diverse surfactants and bile salts without ultrasonication	972–1710 nm	Surfactants studied included SDS, TDOC, DIOCT, DDBS, CPC1	[53]

Technique	Sample	Dispersion method	Measurement	Notes	Ref.
UV-Vis	SWCBT (CoMoCAT and HiPCO)	CNTs were dispersed in different solutions of diverse surfactants and bile salts without ultrasonication	Resonance ratio and normalized width between 539 and 628 nm	Surfactants studied included NaDDBS, SDS, CTAT, CTAB, Triton X-100 and others	[54]
UV-Vis	SWCNT(CoMo CAT and LVD)	CNTs were dispersed in 2% (wt/vol) surfactant solution followed by ultrasonication (40–100 min, 150 W)	Resonance ratio between 300 and 600 nm	Surfactants used are SDS and SDC	[34]
UV-Vis	SWCNT (HiPCo and AD)	CNTs were dispersed in 1 wt% SDS solution followed by ultrasonication (≤0–100 min, 150 W)	Resonance ratio between 300 and 600 nm	Surfactants used are SDS and SDC	[35]
UV-Vis	MWCNT	CNTs were dispersed in 1% wt SDS solution followed by ultrasonication (0–120 min, 20 W)	260 nm	—	[55]
NIR	SWCNT	SWCNTs were dispersed in DMF with ultrasonication and mechanical stirring (5min)	$7500–17,000\ cm^{-1}$	S22 transition peak was used to evaluate the carbonaceous purity of SWCNT bulk samples	[56]
NIRF	SWCNT (CoMoCAT)	CNTs were dispersed in diverse surfactants and dispersing agent solutions followed by ultrasonication (2h,1.5W mL^{-1})	Excitation: 638, 691, and 782 nm emission: 900–1300 nm	Surfactants and dispersing agents included oligonucleotides, peptides, lignin, chitosan, cellulose, and surfactants such as cholates, ionic liquids and organosulfates	[56]
NIRF	SWCNT (CoMoCAT)	SWCNTs were suspended in 2% (wt/vol) solutions of diverse surfactants in deionized water followed by ultrasonication (10 min, 50W)	Excitation:638, 691, and 782 nm; emission: 900–1300 nm	Surfactants and dispersing agents included SDC, BAC, SDBS, SDS, CTAB, Gum Arabic, SC, and Triton X-100	[57]

theoretical values for SWCNTs and MWCNTs were calculated with Debye Waller equation and successfully compared with the experimental neutron diffraction values, which are quite different and influenced by the size, diameter, and interatomic distances of the neighboring atoms. If the diameter is small among the nanoforms, one can easily distinguish them [19]. In addition, we have correlated all the methods related to sample preparation and the characterization data of different spectroscopic analyses are summarized in Table 2.5.

2.3 Conclusion

We have addressed the significant characterization techniques for determining the various forms of CNTs. Although the CNTs are used in so many fields, still there are so many unexplored areas that have potential applications in biology, aerospace, nanoreactors, composite reinforcements, automobile, catalyst supports, probes, and sensors, which need to be explored by the young researchers.

References

1. Ebbesen, T.; Ajayan, P. Large scale synthesis of carbon nanotubes. *Nature*. 1992, 358, 220–222.
2. Oberlin, A.; Endo, M.; Koyama, T. Filamentous growth of carbon through benzene decomposition. *J Cryst. Growth* 1976, 32, 335–349.
3. Iijima, S. Helical microtubules of graphitic carbon. *Nature*. 1991, 56, 354.
4. Prasek, J.; Drbohlavova, J.; Chomoucka, J.; Hubalek, J.; Jasek, O.; Adam, V. Methods for carbon nanotubes synthesis. *J. Mater. Chem.* 2011, 21, 15872–15884.
5. Nag, A.; Mitra, A.; Mukhopadhyay, S.C. Graphene and its sensor based applications: a review. *Sens. Actuators A. Phys.* 2017, 270, 177–194.
6. Kim, P.; Shi, L.; Majumdar, A.; Mc Euen, P.L. Thermal transport measurement of individual multiwalled nanotubes. *Phys. Rev. Lett.* 2001, 87, 215–502.
7. Baughman, R.H.; Zakhidov, A.A.; De Heer, W.A. Carbon nanotubes the route toward applications. *Science* 2002, 297, 787–792.
8. Han, T.; Nag, A.; Mukhopadhyay, S.C.; Xu, Y. Carbon nanotubes and its gas sensing applications: a review. *Sens. Actuators A.* 2019, 291, 107–143.
9. Zuttel, A.; Sudana, P.; Mauron, P.; Kyobayashi, T.; Emenenrgger, C.; Schlapbach, L. Hydrogen storage in carbon nanostructure. *International Journal of Hydrogen Energy*. 2002, 27, 203–212.
10. De Volder, M.F.L.; Tawfick, S.H.; Baughman, R.H.; Hart, A.J. Carbon nanotubes: present and future commercial applications. *Science*. 2013, 339, 535–539.
11. Hu, H.; Bhowik, P.; Zhao, B.; Hamon, M.A.; Itkis, M.E.; Haddon, R.C. Determination of the acidic sites of purified single-walled carbon nanotubes by acid base titration. *Chem. Phys. Lett.* 2001, 345, 25–28.

12. Latorre, C.H.; Mendez, J.A.; Garcia, J.B.; Martin, S.G.; Crecente, R.M.P. Characterization of carbon nanotubes and analytical methods for their determination in environmental and biological samples: a review. *Anal. Chim. Acta* 2015, 853, 77–94.
13. Lee, Y.; Cho, T.; Lee, B.; Rho, J.; An, K.; Lee, Y.J. Surface properties of fluorinated single-walled carbon nanotubes. *J. Flour. Chem.* 2003, 120, 99–104.
14. Gommes, C.; Blacher, S.; Masenelli-Varlot, K.; Bossuot, C.; Mc Rae, E.; Fonseca, A.; Nagy, J.B.; Pirad, J.P. Image analysis characterization of multi-walled carbon nanotubes. *Carbon.* 2003, 41, 2561–2572.
15. Ouyang, M.; Huang, J.L.; Lieber, C.M. Fundamental electronic properties and applications of single-walled carbon nanotubes. *Acc. Chem. Res.* 2002, 35, 12, 1018–1025.
16. Wildoer, J.; Venema, L.; Rinzler, A.; Smalley, R.; Dekker, C. Electronic structure of atomically resolved carbon nanotubes. *Nature* 1998, 391, 59–62.
17. Krasheninnikov, A. Predicted scanning tunneling microscopy images of carbon nanotubes with atomic vacancies. *Solid State Commun.* 2001, 118, 361–365.
18. Cao, A.; Mc Rae, E.; Heyd, R.; Charlier, M.; Moretti, D. Classification for double-walled carbon nanotubes. *Carbon* 1999, 37, 1779–1783.
19. Zhu, W.; Miser, D.; Chan, W.; Hajaligol, M. Characterization of multiwalled carbon nanotubes prepared by carbon arc cathode deposit. *Mater. Chem. Phys.* 2003, 82, 638–647.
20. Saravanan, M.S.; Babu, S.P.; Sivaprasad, K.; Jagannatham, M. Techno-economics of carbon nanotubes produced by open air arc discharge method. *Int. J. Eng. Sci. Technol.* 2010, 2, 5, 100–108.
21. Meenakshi, G.; Nikhil, G.; Kaur, H.; Ashmeet, G.; Kriti, M.; Prashant, J. Fabrication and characterisation of low density polyethylene (LDPE)/multi walled carbon nanotubes (MWCNTs) nano-composites. *Persp. Sci.* 2016, 8, 403–405.
22. Jindal, P.; Goyal, M.; Kumar, N. Modeling composites of multi-walled carbon nanotubes in polycarbonate. *Int. J. Comput. Methods Eng. Sci. Mech.* 2013, 14, 6, 542–551.
23. Jorio, M.A.; Pimenta, A.G.S.; Filho, R.; Saito, G.; Dresselhaus, M.S. Characterizing carbon nanotube samples with resonance Raman scattering. *New J. Phys.* 2003, 5, 139.01–139.17.
24. Hoa, L.T.M. Characterization of multi-walled carbon nanotubes functionalized by a mixture of HNO_3/H_2SO_4. *Diamond and Related Materials.* 2018, 89, 43–51.
25. Suzuki, S.; Hibino, H. Characterization of doped single wall carbon nanotubes by Raman spectroscopy. *Carbon.* 2011, 49, 2264–2272.
26. Li, W.; Zhang, H.; Wang, C.; Zhang, Y.; Xu, L.; Zhu, K.; Xie, S.Raman characterization of aligned carbon nanotubes produced by thermal decomposition of hydrocarbon vapour. *Appl. Phys. Lett.* 1997, 70, 2684–2686.
27. Lopez Lorente, A.I.; Simonet, B.M.; Varcarcel, M. Raman characterization of aligned carbon nanotubes produced by thermal decomposition of hydrocarbon vapour. *Appl. Phys. Lett.* 1997, 70, 2684–2686.
28. Jorio, A.; Satio, R.; Hafner, J.H.; Lieber, C.M.; Hunter, M.; McClure, T.; Dresselhaus, G.; Dresselhaus, M.S. Structural (n, m) determination of isolated single wall carbon nanotubes: influence of the sample aggregation state. *Analyst.* 2014, 139, 290–298.
29. Rao, A.M.; Richter, E.; Bandow, S.; Chase, B.; Eklund, P.C.; Williams, K.A. Diameter selective Raman scattering from vibrational modes in carbon nanotubes. *Science.* 1997, 275, 5297, 187–191.

30. Maultzsch, J.; Reich, S.; Thomsen, C. Chirality selective Raman scattering of the D mode in carbon nanotubes. *Phys. Rev. B.* 2001, 64, 12, 1214071–1214074.
31. Lehman, J.H.; Terrones, M.; Mansfield, E.; Hurst, K.E.; Meunier, V. Evaluating the characteristics of multiwall carbon nanotubes. *Carbon.* 2011, 49, 2581–2602.
32. Belin, T.; Epron, F. Characterization methods of carbon nanotubes: a review. *Mater. Sci. Eng. B.* 2005, 119, 105–118.
33. Kaio, A.B.P.; Sibele, P.C.; Roberto, P.C.N.; Katharina, R.M.M.; Luis, C.M. Oxidized-sulfonated multi-walled carbon nanotube/hydroxyapatite hybrid particles: synthesis and characterization. *J. Solid State Chemistry* 2019, 279, 120924.
34. Yang, B.; Ren, L.; Li, L.; Shi, Y.; Zheng, Y. The characterization of the concentration of single walled carbon nanotubes in aqueous dispersion by UV Vis NIR absorption spectroscopy. *Analyst* 2013, 138, 6671–6676.
35. Yu, J.; Grossiord, N.; Koning, C.E.; Loos, J. Controlling the dispersion of multiwall carbon nanotubes in aqueous surfactant solution. *Carbon.* 2007, 45, 618–623.
36. Zeisler, R.; Oflaz, R.; Paul, R.L.; Fagan, J.A. Use of neutron activation analysis for the characterization of single walled carbon nanotube materials. *J. Radioanal. Nucl. Chem.* 2012, 291, 561–567.
37. Pang, L.S.K.; Saxby, J.D.; Chatfield, S.P. Thermogravimetric analysis of carbon nanotubes and nanoparticles. *J. Phys. Chem.* 1993, 97, 6941–6942.
38. Lima, A.; Musumeci, A.; Liu, H.W.; Waclawik, E.; Siva, G. Purity evaluation and influence of carbon nanotube on carbon nanotube/graphite thermal stability. *J. Therm. Anal. Calorim.* 2009, 97, 1, 257–263.
39. Dunens, O.M.; Mackenzie, K.J.; Harris, A.T. Synthesis of multiwalled carbon-nanotube on fly ash derived catalysts. *Environ. Sci. Technol.* 2009, 43, 20, 7889–7894.
40. Scheibe, B.; BorowiakPalen, E.; Kalenczuk, R.J. Oxidation and reduction of multiwalled carbon nanotubes-preparation and characterization. *Mater. Charact.* 2010, 61, 2, 185–191.
41. Kim, D.Y.; Yang, C.M.; Park, Y.S.; Kim, K.K.; Jeong, S.Y.; Han, J.H. Characterization of thin multiwalled carbon nanotubes synthesized by catalytic chemical vapor deposition. *Chem. Phys. Lett.* 2005, 413, 1–3, 135–141.
42. Don Young, K.; Young Soo, Y.; Soon Min, K.; Hyoung Joon, J. Preparation of aspect ratio controlled carbon nanotubes. *Mol. Cryst. Liq. Cryst.* 2009, 510, 79–86.
43. Mawhinney, D.B.; Naumenko, V.; Kuznetsov, A.; Yates, J.T., Jr.; Liu, J.; Smalley, R.E. Optimization of Xe adsorption kinetics in single walled carbon nanotubes. *Chem. Phys. Lett.* 2000, 324, 213–216.
44. Burian, A.; Koloczk, J.; Dore, J.; Hannon, A.C.; Nagy, J.B.; Fonseca, A. Radial distribution function analysis of spatial atomic correlations in carbon nanotubes. *Diam. Relat. Mater.* 2004, 13, 1261–1265.
45. Dore, J.C.; Sliwinski, M.; Burian, A.; Howells, W.S.; Cazorla, D. Structural studies of activated carbons by pulsed neutron diffraction. *J. Phys. Condens. Mater.* 1999, 11, 9189.
46. Giannasi, A.; Celli, M.; Sauvajol, J.; Zoppi, M.; Bowron, D. SWCN characterization by neutron diffraction. *Physica B* 2004, 350, E1027–E1029.
47. Bokobza, L.; Zhang, J. Raman spectroscopic characterization of multiwall carbon nanotubes and of composites. *Express Polym. Lett.* 2012, 6, 601–608.
48. Lopez Lorente, A.I.; Simonet, B.M.; Varcarcel, M. Raman spectroscopic characterization of single walled carbon nanotubes: influence of the sample aggregation state. *Analyst* 2014, 139, 290–298.

49. Li, R.; Chang, X.; Li, Z.; Zang, Z.; Hu, Z.; Li, D.; Tu, Z. Multiwalled carbon nanotubes modified with 2-aminobenzothiazole modified for unique selective solid phase extraction and determination of Pb(II) ion in water samples. *Microchim. Acta* 2011, 172, 269–276.

50. Zang, Z.; Hu, Z.; Li, Z.; He, Q.; Chang, X. Synthesis, characterization and application of ethylenediamine modified multi-walled carbon nanotubes for selective solid phase extraction and preconcentration of metal ions. *J. Hazard. Mater.* 2009, 172, 958–963.

51. Meng, J.; Yang, M.; Song, L.; Kong, H.; Wang, C.Y.; Wang, R.; Wang, C.; Xie, S. S.; Xu, H.Y. Concentration control of carbon nanotubes in aqueous solution and its influence on the growth behaviour of fibroblasts. *Colloid Surface* 2009, 71, 148–153.

52. Attal, S.; Thiruvengadathan, R.; Regev, O. Determination of the concentration of single walled carbon nanotubes in aqueous dispersion using UV-Vis absorption spectroscopy. *Anal. Chem.* 2006, 78, 8098–8104.

53. Wensellers, W.; Vlasov, I.I.; Goovaerts, E.; Obraztsova, E.D.; Lobach, A.S.; Bouwen, A. Efficient isolation and solubilization of pristine single walled nanotubes in bile salt micelles. *Adv. Func. Mater.* 2004, 14, 1105–1112.

54. Tan, Y.; Resasco, D.E. Dispersion of single walled carbon nanotubes of narrow diameter distribution. *J. Phys. Lett.* 2000, 317, 497–503.

55. Itkis, M.E.; Perea, D.E.; Niyogi, S.; Rickard, S.M.; Hamon, M.A.; Hu, H.; Zhao, B.; Haddon, R.C. Purity evaluation of as prepared single walled carbon nanotube soot by use of solution phase near IR spectroscopy. *Nano Lett.* 2003, 3, 309–314.

56. Haggenmueller, R.; Rahatekar, S.S.; Fagan, J.A.; Chun, J.; Becker, M.L.; Naik, T.; Krauss, R.R.; Carlson, L.; Kadla, J.F.; Trulove, P.C.; Fox, D.F.; Delong, H.C.; Fang, Z.; Kelley, S.O.; Gilman, J.W. Comparison of the quality of aqueous dispersions of single wall carbon nanotubes using surfactants and biomolecules. *Langmuir.* 2008, 24, 5070–5078.

57. Schierz, A.; Parks, A.N.; Washburn, K.M.; Chandler, G.T.; Ferguson, P.L. Characterization and quantitative analysis of single walled carbon nanotubes. *Environ. Sci. Technol.* 2012, 46, 12262–12271.

Author Information

Sannapaneni Janardan is Assistant Professor at the Department of Chemistry, School of Sciences of the GITAM (deemed to be University), Bengaluru. He has received PhD in organometallic chemistry from the VIT (deemed to be University) in the year 2016. He has experience in diverse range of research areas, which include organometallics, material science, and bioorganic chemistry. His areas of research interests are organometallics, materials, and bioorganic chemistry.

Acronyms

CNT	Carbon nanotubes
oCNT	Oxidized multi walled carbon nanotubes
HAP	Hydroxy apatite
oCNT- HAP	oxidized multiwalled carbon nanotubes- Hydroxyapatite
SDS	Sodium dodecyl sulfate
CTAT	hexadecyltrimethyl ammonium *p*-toulenesulfonate
CTAB	Cetyl trimethyl ammonium bromide
NaDDBS	Sodium dodecyl benzenesulfonate
SC	Sodium chlorate
TDOC	Sodium taurodeoxy cholate
DIOCT	Dioctyl sulfosuccinate
DDBS	Dodecyl benzene sulfonate
CPC1	Cetylpyridinium chloride
CoMoCAT	cobalt molybdenum catalysis
HiPCO	High-pressure carbonmonoxide SWCNTs
DTA	Differential thermal analysis
FT-IR	Fourier transform infrared spectroscopy
MWCNT	Multiwalled carbon nanotubes
NIR	Near infrared
NIRF	Near infrared spectroscopy
RS	Raman spectroscopy
SEM	Scanning electron microscopy
STM	Scanning tunneling microscopy
SWCNT	Single walled carbon nanotubes
TEM	Tunneling electron microscopy
TGA	Thermogravimetric analysis
UV-Vis	Ultraviolet Visible Spectroscopy
XPS	X-ray photoelectron spectroscopy
XRD	X-ray diffraction
EDS	Energy-dispersive spectroscopy
SEC	Size-exclusion chromatography
AFM	Atomic Force Microscopy
ND	Neutron diffraction

3

Influence of Gulliver over Lilliput: Novel Properties of Carbon Nanotube/Liquid Crystal Composites

S. Krishna Prasad, D.S. Shankar Rao, Marlin Baral, and G.V. Varshini
Centre for Nano and Soft Matter Sciences
Bengaluru, India

3.1 Introduction

Over the past decades from different perspectives, two well-known materials, namely carbon nanotubes (CNTs) and liquid crystals (LCs) have established themselves for significant applications. As materials, these two categories are widely far apart but indeed a few similarities can be found, owing to which composites of these two materials are actively pursued for synergistic aspects such as, for example, combination of high electrical conductivity of CNTs and fluid, yet anisotropic properties of LCs.[1–4] In such hybrid systems, LC finds its role for the ability of its molecules to get oriented by surface forces or moderate external fields.[5] In composites of LC and CNT, a further advantage is that even in the absence of a direct influence of the external stimuli on the CNT, owing to sympathetic elastic causes, the orientation of the nanotubes could be governed by that of the LC molecules. A proper realization of such a feature has the potential to result in devices in which macroscopic properties can be switched between their anisotropic values along, say, parallel to perpendicular direction with respect to an internal reference axis. In fact, reorientation of the nematic director from the equilibrium direction to that dictated by the field, and the concomitant change in the parameters such as electrical conductivity caused by the presence of CNTs, have been well demonstrated. In this chapter, we have considered three case studies carried out in our lab to highlight the advantages of the CNT/LC composites.

In the following sections, we provide a brief introduction to, in general, LCs and, in specific, to two liquid crystalline phases, which will be discussed in this chapter. LCs are states of matter that simultaneously have properties of a liquid, such as fluidity, and of crystals with characteristics such as anisotropy in dielectric and optical properties. Exhaustive descriptions of the structure and physical properties of LCs can be found in standard text books.[6] Depending upon the shape anisotropy, thermotropic LC (wherein liquid crystallinity is brought about by the action of heat) molecules are generally rod-like (calamitic) or disk-like (discotic); in recent times banana-shaped (bent-core) molecules have been attracting much attention (see Figure 3.1). One of the conditions to favor the liquid crystallinity is the aspect ratio of the length to diameter of the rod-like (also banana-shaped) mesogens or the ratio of the diameter to the thickness of the disk-shaped molecules, being five or above. With hardly an exception, calamitic mesogens comprise a rigid core, often aromatic, to which one or more flexible alkyl chains are attached. Such mesogens do give rise to as many as 40 different LC structures, but here we limit our description to only two of them, namely, nematic and smectic A (SmA), which are discussed in this chapter. In the simplest of all the LCs, the nematic (N) phase exhibits long-range orientational order but no positional order. The orientational order is described in terms of the ability of the molecules to be oriented in a particular direction, referred to as the director (**n**), which is apolar, **n** = −**n**. The smectic phases are characterized by a layered structure besides the orientational order of the N phase. Based on the molecular arrangements within and between the layers, smectic mesophases are classified into different types. In the SmA phase, the director is collinear with the normal to the layer planes with the centers of mass of the molecules irregularly placed within the layer in a liquid-like manner. The layering arrangement is more rigorously treated in terms of a one-dimensional mass density wave.

The liquid crystalline medium exhibits anisotropic character in several physical properties, such as optical, electric, and magnetic. For example, the N and SmA phases show two refractive indices, labeled, ordinary (n_o,

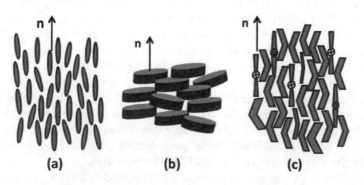

(a)   **(b)**   **(c)**

FIGURE 3.1
Schematic arrangement of the molecular arrangement in the nematic phase formed by (a) calamatic, (b) disk-shaped, and (c) banana-shaped molecule.

perpendicular to the director) and extraordinary (n_e, along the director) refractive indices. The birefringence of the medium, given by the difference $\Delta n = n_e - n_o$, is thus finite, and generally for rod-like systems is positive. In a similar fashion, the dielectric and diamagnetic anisotropies are defined as $\Delta\varepsilon = \varepsilon_\| - \varepsilon_\perp$, and $\Delta\chi = \chi_\| - \chi_\perp$, where ‖ and ⊥ indicate directions parallel and perpendicular to the director, respectively, are positive.

This chapter gives exemplary illustrations of three different aspects wherein the presence of CNT, albeit in small concentrations, has a substantial influence on the LC property. The cases presented are (i) the thermal variation and field-driven switching of anisotropic electrical conductivity, (ii) reinforcement of the polymer network in a polymer-stabilized LC system, and (iii) induction/stabilization of a layered phase in a binary mixture exhibiting a re-entrant phase sequence.

3.2 Materials and Methods

The different liquid crystalline host materials employed for preparing the CNT/LC composites are listed out in Table 3.1 along with the transition temperatures exhibited by them. In all the composites, the CNT composition was kept at low values, a requirement to ensure that there is no macroscopic separation of the two materials; the highest concentration of CNT employed was 0.5% by weight. The CNTs employed were single-wall carbon nanotubes (SWCNTs from Heji, Hong Kong) with a nominal purity of >90%, diameter of about 2 nm with a peak length of 500 nm, providing an aspect ratio of up to 250.

Uniform alignment of CNTs is essential, and in general the ability of providing nanotubes a predetermined direction is of great importance. Additionally, it is desirable to be able to manipulate this direction, for example, by the application of external fields, such as electric, magnetic, or mechanical. The electric field process of reorienting LCs, known as the Freedericksz transition,[7] employed in most common LCD applications, has been attempted[5,8] to get sympathetic reorientation of dispersed CNTs.

3.2.1 Preparation of the Composites

A challenging, but an essential requirement, for the studies was to have good dispersions of CNTs in the LC medium.[3] The strong van der Waals interaction between neighboring nanotubes leads to aggregation and even possible phase separation of the two constituents. The prepared composite also appears homogeneous but most CNTs show sedimentation within a few days. High-power ultrasonication for prolonged times while keeping the tip sonicator in the material, may improve the situation to some extent.[9,10] However, much simpler method providing better long-term

TABLE 3.1

Liquid crystalline materials employed along with their transition temperatures

Short Form	Structure and Transition Temperatures (°C)
PCPBB	
	Iso 120 °C N
E7	Eutectic mixture from E Merck
	CN 5CB (51%)
	CN 7CB (25%)
	CN 8OCB (16%)
	CN 5CT (8%)
	Iso 64 N
60CB	
	Iso 76.4 N
8OCB	
	Iso 80.2 N 67 SmA

stability just involves stirring the sample (LC+CNT) gently with a magnetic stir bar for a few hours. For this purpose, weighed amounts of the two components contained in a glass vial with a volatile solvent were heated to high temperature, subsequent to which the mixture was subjected to magnetic stirring for ~15 h, cooled to room temperature with the stirring continued till the samples were filled into the test cells. The uniformity of the dispersions could be seen visually and further confirmed by observation of a sharp clearing point (the nematic-isotropic transition temperature). It may be mentioned that, as pointed out by Lagerwall et al.,[11] the stirring or the ultrasonication method does not ensure bundle-free mixtures, but keeps the bundle size relatively small. As seen under polarizing optical microscope, the nanoparticles are quite randomly dispersed with not too large a size at any point. This could be the cause for the good uniformity of the gently stirred LC+CNT composites.

3.3 Results and Discussion

3.3.1 Electrical Conductivity

3.3.1.1 Thermal Variation

The electrical conductivity is a significant property of the material being governed by the nature of charge carriers, such as electron/holes or cations/anions, and investigated for as a function of temperature and frequency to elicit information about the underlying mechanism. The temperature dependences of the low-frequency (100 Hz) anisotropic electrical conductivities, parallel ($\sigma_\parallel$) and perpendicular ($\sigma_\perp$) to the director, for the pure LC (PCPBB), and its composite with CNT (PCPBB+0.05% CNT) obtained using a small probing voltage are shown in Figure 3.2(a). The first feature to be noticed is that upon addition of CNT, both $\sigma_\parallel$ and $\sigma_\perp$ increase by about two orders of magnitude, and that the charge transport is still anisotropic; as expected, the $\sigma_\parallel$ and $\sigma_\perp$ values merge at the onset of T_{NI}. The more interesting feature, however, is the drastically different thermal dependences of σ for the two samples. While pure LC exhibits, the expected stronger variation with the value decreasing by nearly two orders of magnitude over the range of the nematic phase, the composite presents very weak temperature dependence. The range covered of >70 K is by far the largest for typical LC phase. It may be pointed out that among the composites of nanoparticle-doped LC systems this feature is unique for the CNT-containing materials.[12,13] For example, LC composites of either gold nanoparticles[14,15] or nanorods[16,17] and even in a gel environment[18] retain the large variation of σ characteristic of the host LC.

The thermal variation data shown in Figure 3.2(a) can be very well described by the Arrhenius equation,

$$\sigma = \sigma_o \exp\left(-E_g/kT\right) \tag{3.1}$$

where σ_o is the pre-exponential factor corresponding to $1/T = 0$, E_g, the conductivity activation energy, and k, the Boltzmann constant; in the

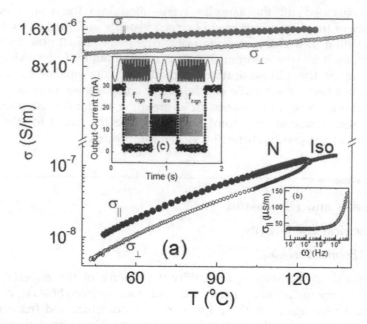

FIGURE 3.2
(a) Thermal variation of the electrical conductivity parallel ($\sigma_\parallel$) and perpendicular ($\sigma_\perp$) to the nematic director, and in the isotropic phase obtained at 100 Hz probing frequency for PCPBB (the data set in the lower portion of the panel) and the PCPBB+0.05% CNT composite (the data set in the upper portion of the panel). The composite has two orders of magnitude higher conductivity and retains the anisotropy, but more importantly, exhibits negligible temperature dependence of the values. (b) The frequency dependence of $\sigma_\parallel$ of PCPBB+0.05% CNT composite. The solid line represents the best fit of the data to Equation (3.2). (c) Switching of the output current (filled circles) of the CNT composite with the driving profile shown as solid line, wherein the frequency (of 10 V magnitude) is changed between f_{high} = 600 kHz and f_{low} = 500 Hz. The switching current (or equivalently conductivity) is due to the change in the orientation direction of the LC molecules, as conformed by simultaneous change of the microscopy field of view from the birefringent to a dark one, the texture as shown in the images (d–f), shown in the respective regions.

hopping scenario, E_g represents the potential barrier for the charge carrier to hop from one site to another. Fitting the data to Equation (3.1), it is found that ΔE decreases drastically from 34.4 ± 0.05 kJ/mol for the pure LC (a value typical for LC materials), to a quite small value of 2.68 ± 0.02 kJ/mol for the CNT composite. Indeed Lebovka et al.[19] observed a similar behavior for another LC+CNT composite, and provided an explanation based on the percolation model, with the LC and LC+CNT composites considered to be, respectively, below and above the system's percolation threshold. The model further expounds that the conductivity behavior of the LC is dominated by ionic transport mechanism, as would be in a semiconductor having predominantly thermally assisted hopping or charge tunneling. In contrast, for the composite, the electronic conductivity of CNT dominates resulting in weaker thermal dependence. A percolation threshold is anticipated, but the concentration could be very low, unlike in metal nanoparticle-LC system wherein such thresholds are clearly established.

In stark contrast with this behavior, the gold nanoparticle-LC composites embedded in a gel environment actually exhibit the super-Arrhenius behavior, commonly seen in glass-forming molecular liquids and polymeric systems. Describing the temperature dependence by the standard Vogel–Fulcher–Tammann expression, we have found fragile glass behavior in these composites.[18] An effective communication between CNTs through percolation network formation and relatively more important electronic conductivity is perhaps responsible for differentiating their behavior in contrast to that seen for systems with other nanoparticles.

3.3.1.2 Frequency Dependence of Conductivity

The frequency spectrum of conductivity is characteristic of conductors formed by particle-doped insulators, and there has been overwhelming evidence for universal behavior of the AC conductivity for a wide range of materials including ion-conducting glasses, perovskites, amorphous and polycrystalline semiconductors, transition metal oxides, and metal cluster compounds. This general behavior can be described by Jonscher's[20] universal dynamic response. The typical behavior at frequencies higher than the region where electrode polarization effects can be seen is a constant conductivity (σ_{DC}) at lower frequencies, and strong frequency dependence (σ_{AC}) above a critical frequency, ω_c. Jonscher's[20] description is based on distribution of hopping probabilities between sites located randomly in space/energy and is mathematically expressed as a power law,

$$\sigma(\omega) = \sigma_{DC} + A\omega^n \tag{3.2}$$

where A is a pre-exponential factor. The all-important exponent is a measure of the interaction between the charge carriers and the environment and in disordered solids, generally takes a value between 0 and 1. Figure 3.2(b) presents the frequency-dependent $\sigma_\perp$ data for the composite. The qualitative behavior conforms to the Jonscher[20] expectation, which is further confirmed by the fitting done to Equation (3.2); the solid line through the data shows that Equation (3.2) describes the experimental situation very well. However, the fitted value of the exponent, $n = 1.058 \pm 0.003$ is higher than the universal expectation. As stated above, the validity of the fractional power law of universal dynamic response is well established in diverse disordered systems and thus the feature $n > 1$ observed in ion-conducting glasses,[21] ionic crystals[22] prompted further theoretical analysis by Papathanassiou et al.[23] The case of CNT-driven conductivity phenomenon is all the more relevant for such analysis owing to the fact that in addition to classical hopping over potential energy barriers, quantum mechanical tunneling of delocalized electrons through these barriers becomes important. Papathanassiou et al.[23] described a model involving conduction paths having distribution of path lengths, which permits situations with $n > 1$ as well. The second parameter that could cause the deviation from $n < 1$ condition is the thermal fluctuations inherent to the nematic state, which might affect the contact between two neighboring CNTs. The substantial increase in conductivity between the LC host and the composite, and the weak thermal variation for the latter, unequivocally show that the system is above the percolation threshold. However, since the concentration of CNTs is still low, the junction resistances between CNTs dispersed in the LC medium can, in the general case, be taken to be larger than the resistance of the CNTs themselves, and thus any possible network of CNTs could be imagined to have randomly distributed barriers for electrical transport. In such a scenario, the length of the conducting path becomes a spatially and temporally varying parameter.

3.3.2 Dual Frequency Switching

The studied host LC, PCPBB, exhibits the dual frequency switching characteristic,[24] which we will briefly explain here. Over the frequency range of interest (<1 MHz), $\varepsilon_\perp$ is weakly dependent on the probing frequency f. In contrast, $\varepsilon_\parallel$ is quite sensitive to f and exhibits a relaxation. At a particular frequency, labeled crossover frequency (f_{cr}), which is related to the relaxation frequency f_R, $\varepsilon_\perp$ cuts through the $\varepsilon_\parallel$ data. This results in the dielectric anisotropy ε_a ($= \varepsilon_\parallel - \varepsilon_\perp$) to be positive below f_{cr}, and negative above it. Thus, with an applied electric field E (above a certain threshold to be discussed in the next section), the dielectric torque $\varepsilon_a E^2$ on the nematic director can be either positive or negative depending on the frequency f of the field being lower or higher than a critical value, f_{cr}. Thus, for materials like PCPBB, the molecules are driven parallel for $f < f_{cr}$ or perpendicular to E for $f > f_{cr}$. Concomitant with such frequency-controlled reorientation of the molecules, the conductivity or current through the system could switch the two anisotropic limits. Figure 3.2(c) demonstrates this feature obtained with a voltage of 10 V while alternating the frequency of the field between 500 Hz ($\ll f_{cr}$) and 600 kHz ($\gg f_{cr}$). The alternation between high- and low-conducting states presents a current ratio of 170:1. This dual-frequency-driven flipping of conductivity can be used as an optical switch as well. The sample assembly is such that in the equilibrium situation (no field), the LC molecules occupy the plane of the substrate, and therefore the system exhibits high birefringence, as seen in the polarizing microscopy image presented in Figure 3.2(d). Application of the low-frequency (500 Hz) field drives the molecules to be normal to the substrate, and owing to the sample being between crossed polarizers, the field of view is dark (Figure 3.2(e)). The return to the equilibrium planar state (Figure 3.2(f)) is achieved by merely switching the frequency of the field to 600 kHz. It may be pointed out that the return to the planar state could, in principle, be obtained by just turning the field off completely, deploying the high-frequency component results in two orders of magnitude faster switching, with the response time in milliseconds. The protocol is quite generic and thus through proper choice of materials holds the potential to achieve the desirable goal of switching between insulating and good semiconducting states.

3.3.3 CNT Reinforced Polymer-Stabilized LCs

From the viewpoint of tunable electro-optic properties, which could lead to applications such as smart glass, composites of polymers and LCs have attracted much attention.[25] The emphasis in such devices is to appeal to equilibrium-scattering and field-driven transparent phenomenon. These devices are based on the anisotropic refractive indices of LC, but are not operated in the birefringent mode, unlike the devices mentioned in the previous section. Thus, they can do away with the polarizers completely, a component in the common LC devices that limits their viewing angle. Contingent on the concentration of the polymer component, two categories exist: polymer-dispersed LCs (PDLC), if the polymer is the major component, and polymer-stabilized LCs (PSLC), if LC is the major component. The preparation involves a local level of phase separation of polymer and LC through thermal, solvent, or optical processes. The polymer is chosen such that its refractive index n_p matches one of the refractive indices, say, the ordinary value, n_o. This feature in conjunction with the ability of LC molecules to reorient upon application of a suitable field and consequent change in their refractive index lies at the heart of the device operation. In the device configuration that will be described in this chapter, in the equilibrium state the LC molecules lie in the plane of the substrate, and thus exhibit the extraordinary refractive index, $n_e \neq n_p$. This condition along with the fabrication controlled droplet size of LC in the case of PDLC, or polymer fibril size in PSLC results in significant scattering of the incident light. When an electric field that couples to the dielectric anisotropy of the LCs is applied, the LC molecules reorient changing their refractive index to n_o. Since $n_o = n_p$, the device becomes completely transparent.

Smart switchable e-glass based on the above-described principle is already commercially available.[25] The drawback of PDLC/PSLC devices, however, is that V_{th}, the minimum operational voltage required to initiate the reorientation from the surface-determined one, has not only a higher value than for the pure LC, but continues to have the same temperature dependence. Among the persistent efforts to improve this situation, concept of having nanoparticles incorporated into the system[26–33] is relevant here. The results of these reports can be summed up as, addition of CNT does not generally lower V_{th} and also no notable reduction in response time is seen. Further, the response–slope (dC/dV, where C is the sample capacitance) gets adversely affected by diminishing. Improvement in each of these features is certainly advantageous for lower operating voltage, faster responding system that would hasten the screen refreshing time, and providing flicker-free images and better gray capability. Attempts having simple physical mixing of CNTs and LCs, wherein the nanotubes are free to diffuse through the medium, especially the fluid LC regions, have not been found beneficial.

3.3.3.1 Materials

With this in the background, we recently[34] undertook investigations that realize proper functionalization of CNTs so as to retain in the polymer matrix itself, with the polymer itself being appropriate for PSLC formation. For this, the surface of the CNTs was decorated with the same polymer that also creates polymer stabilization, a feature that prevents CNTs from diffusing away from the polymer areas, and presumably also strengthening the polymer fibrils.

The LC used is the commercially available eutectic mixture E7. The polymer matrix was formed by polymethyl methacrylate (PMMA). The novel ingredient employed is the modified PMMA having the PMMA polymer strands grafted on to SWCNTs (schematically shown in Figure 3.3(a)) having an aspect ratio >1000. In the studies described here, the LC: polymer (bare PMMA as well as the grafted one) were in the ratio of 95:5, by weight. Before mixing, the LC and polymer were first dispersed separately in chloroform, mixed, and then sonicated followed by removal of the solvent at low vacuum for extended duration. The composite was then maintained at 150 °C over a short time and subsequently cooled to room temperature. Hereafter, the composite with only PMMA, and that with CNT are referred to as PMMA-b and PMMA-CNT, respectively.

3.3.3.2 Morphological Features

Before discussing the electro-optic characteristics, let us look at the morphological difference that the CNT brings about. A schematic representation of the PSLC architecture is shown in Figure 3.3(b), the polymeric material forms fibrils and a scaffold in which the LC gets confined; this simple case being applicable to the PMMA-b sample. On the other hand, the PSLC architecture in the PMMA-CNT case can be visualized, as depicted in Figure 3.3(c), to have at least in some of the regular PMMA fibrils, the CNT-tethered PMMA. In fact, the SEM image obtained from the PSLC cell after leaching out the LC exhibits the expected collapsed fibrillar morphology for PMMA-b (Figure 3.3(d)). In contrast, even under the optical microscope, the PMMA-CNT sample presents the *Swiss-cheese* architecture, a feature confirmed by SEM imaging (Figure 3.3

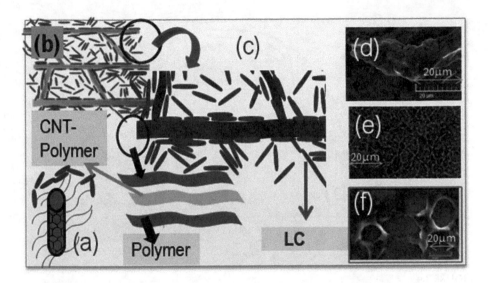

FIGURE 3.3
(a) Schematic representation of PMMA-tethered CNT in the vicinity of liquid crystal (material: E7) molecules shown as ellipses. (b) PSLC architecture wherein the LC molecules (ellipses) get confined in the polymer network (stripes). (c) In the case of the PSLC containing the grafted CNT, the polymer network gets strengthened by the CNT. LC molecules lying next to the polymer fibrils will have their orientation dictated by the polymer walls. In PSLCs with PMMA-b, the network is of the fibrillar type as shown by SEM imaging (d). In contrast, PSLC having the tethered CNT as well exhibits Swiss-cheese architecture as seen in the optical microscopy (e) and SEM (f) images.

(e) and (f), respectively). This difference in the morphology suggests that the CNT-tethered PMMA present inside the PMMA fibrils in fact provides a mechanical reinforcement leading to stronger polymer network that sustains collapse during the leaching out process. The *Swiss-cheese* architecture resembling the droplet morphology is a distinctive characteristic of polymer-dispersed LCs, in which, as mentioned earlier, the polymer is the major constituent that necessarily improves the rigidity of the device. Thus, the embedding of the tethered CNT in PMMA seems to achieve, with much smaller polymer content, the same mechanical robustness of PDLC. The stronger fibers could be expected to enhance the anchoring strength of LC molecules at the polymer interfaces, a feature that is perhaps responsible for the novel electro-optic characteristics described in the next section.

3.3.3.3 Electro-optics

The Freedericksz transformation behavior for the two PMMA-b and PMMA-CNT systems is presented in Figure 3.4(a). The configuration employed is such

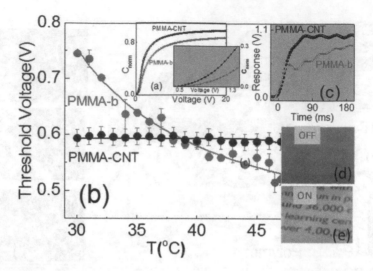

FIGURE 3.4

(a) Freedericksz transformation profiles presenting voltage dependence of $C_{norm} = (C - C_\perp)/C_\perp$, with C being the sample capacitance having a value $C_\perp$ in the equilibrium planar state. The composite shows a lower threshold (V_{th}) as well as a higher slope post threshold, evident in the data presented on an enlarged scale, in the inset. (b) Drastically different temperature dependence of V_{th} for the no-CNT (PMMA-b) and with CNT (PMMA-CNT) PSLC architectures, with the latter being thermally invariant. The superior characteristics extend to the dynamic characteristics, especially the switch-off process obtained under crossed polarizers as shown in (c). Note that the response is through two steps for PMMA-b, but only one for PMMA-CNT, resulting in a faster response. The PMMA-CNT electro-optic device in the (d) field-off scattering and (e) field-on transparent states presenting significant contrast for viewing text kept behind the device.

that in the equilibrium state with no applied voltage, the LC molecules lie in the glass substrate (and electrode) plane. When the voltage exceeds a certain critical value, known as the Freedericksz threshold (V_{th}), the molecules start reorienting toward the electric field, and at much higher voltages become normal to the substrate. Since this process of planar to homeotropic orientation change is also associated with a change in the permittivity of the medium from the equilibrium $\varepsilon_\perp$ to the high-field $\varepsilon_\parallel$, it can be easily tracked by capacitance measurements. In Figure 3.4(a), such data are, for presentation purposes, shown after normalizing with respect to the equilibrium capacitance value $(C_\perp)$. It is evident that the CNT composite has a lower V_{th} than the PSLC mixture without CNT (PMMA-b). Further, the rise in capacitance above V_{th} is sharper for the PMMA-CNT sample, a behavior related to the bend elastic constant, is an attractive feature for better gray scale capability of the device; quantified in terms of the dC/dV, this parameter is doubled for the PSLC with CNT.

The most prominent improvement realized by the presence of CNT in the polymer strands is in the temperature dependence of the threshold voltage (Figure 3.4(b)).While PMMA-b device exhibits the expected strong increase on lowering the temperature, the PMMA-CNT composite, even at the low concentration of the PMMA decorated CNT, an essential temperature independence of V_{th}. This feature gains significance for designing display driving circuits as it obviates the necessity to take into account the device operating temperature variation. The dynamic response of the PMMA-CNT device also presents advantages over its non-CNT counterpart, especially the switch-off response. As shown by the temporal variation of the intensity transmitted through the device when the applied voltage is turned off (Figure 3.4(c)), the PMMA-CNT device returns much faster (by a factor of 2) to the equilibrium value, a feature highly repeatable. This acceleration is true over the entire temperature range of the nematic, spanning 25 K. These features are found in the scattering/transmission as well as birefringent modes (obtained by placing the device between crossed polarizers). Besides this field-off response is governed by a single process for PMMA-CNT unlike the two-time scales, the second of which being much longer, for the PMMA-b case. The single response certainly makes the addressing protocols much simpler. The operation of the device in the scattering and transmission states is shown as images, respectively, in Figure 3.4(d) and (e). The contrast between the field-off scattering and field-on transparent states for viewing printed text placed behind the device is significantly high.

3.3.3.4 Gas Sensor

Now we describe a simple gas sensor based on the PMMA-CNT device, which is under development. Unlike the usual strategy employed,[35] this does not involve forced flow of volatile organic compounds (VOCs), but the natural evaporation process, and thus mimics real life situations. A schematic diagram of this gas sensing device is shown in Figure 3.5(a). It consists of a glass reservoir of 10 ml volume, which is covered with an impermeable lid having two orifices: the larger of the two, 5 mm in diameter, matches the dimensions of the sample placed over it. The second, much smaller, is used to introduce the VOC using the needle of a syringe, but kept closed otherwise. The leads from the electrodes sandwiching the sample are connected to an impedance analyzer (HP 4194A) employed to measure AC conductance of the sample at 1 kHz. A typical experiment involved collecting the baseline data in the absence of VOC, introduce the VOC into the reservoir through the syringe, obtain data and finally monitor the decay in the response by shutting up the VOC supply. The complete response curve over one cycle, in terms of temporal variation of conductance, obtained with acetone as the VOC is shown in Figure 3.5(b). On getting the VOC, the device conductance enhances by a factor of 3 with a quick rise followed by a slower increase toward a limiting value. The relaxation to the equilibrium value subsequent to stopping the flow takes a much longer duration. The device has not only good responsivity but also has good selectivity, as shown in Figure 3.5(c) comparing the response curve for four different VOCs, namely acetone, ethanol,

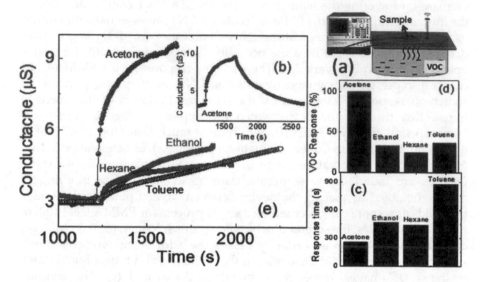

FIGURE 3.5

(a) Schematic representation of the gas sensor device. (b) Conductance variation of the PMMA-CNT device through full cycle of exposure to acetone vapors and return to the equilibrium. (e) Temporal behavior of conductance upon exposure to different VOCs, exhibiting detection as well as selectivity. (d) Comparison among the four VOCs used in terms of the conductance variation (see text for definition of VOC response%). (c) Response time of the device when exposed to different VOCs.

hexane, and toluene. The magnitude of the response, in terms of the VOC response % = $(S_s - S_c)/S_c$, where S_s and S_c are the saturated and baseline conductances, respectively, shown in Figure 3.5(d) demonstrates that the device is most sensitive to acetone. In the light of experimental difficulties to carry out the studies with non-VOCs, the response for hexane could itself be taken to contain external influences, even if present. With this caveat, acetone certainly has a specific response. Similarly, the response time, taken as the duration for the response to vary from 10% to 90% of the total variation, displayed in Figure 3.5(e) for the different VOCs, is the fastest for acetone. Thus, this simple device demonstrates the excellent potential of PMMA-CNT composites for efficient detection of acetone. Investigations are underway to fabricate the device on substrates with interdigitated electrodes and usage of apparatus with mass flow controllers is underway and expected to result in better characterization of the device.

3.3.4 Phase Induction/Stabilization

In this last section, we describe induction of a layered phase or the enhancement of its thermal range by the doped CNT, again at very low concentrations. The well-known binary phase diagram of the host LC system comprising the 6th and 8th homologues of the *n*-alkoxy cynao biphenyl series (6OCB and 8OCB, respectively) is shown in Figure 3.6(a). The feature of specific interest, from the viewpoint of present studies is the occurrence of the re-entrant sequence over a range of concentrations. Generally, when a material is cooled, the sequence of phases observed is such that a disordered or high symmetry phase is followed by a more ordered or lower symmetry one. However, in certain cases, due to competing interactions, the less ordered phase could again appear at a lower temperature, the phenomenon being referred to as re-entrancy.[36] A wide variety of systems ranging from LCs[37,38] binary fluids,[39] ferroelectrics,[40] magnets,[41] superconductors,[42] proteins,[43] and even as exotic as astrophysical phenomenon of black holes[44] have been reported to exhibit re-entrance. Plethora of re-entrant sequences has been observed in different LC systems[45,46] with the nematic-SmA-re-entrant nematic seen in binary mixtures of strongly polar materials, such as the one shown in Figure 3.6(a), being one of the first.[47]

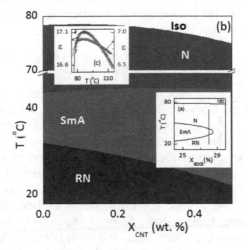

FIGURE 3.6
(a) Partial phase diagram of the 6OCB/8OCB binary system as a function of the wt% of 6OCB. Studies described in this chapter have been carried out on two host mixtures whose concentrations are shown as vertical lines (solid and dotted lines represent compositions 28% and 29.4% of 6OCB in 8OCB, respectively). (b) Effect of CNT composition on the different boundaries of the host LC mixture with 28% of 6OCB in 8OCB. The salient feature is that the SmA-RN boundary is significantly lowered in temperature and concomitantly the thermal range of the SmA range is enhanced. (c) Thermal variation of the permittivity on approaching the nematic from the isotropic phase in the host LC mixture and the $X = 0.24$ CNT composite, exhibiting convex-shaped anomaly. Arrows (upward and downward) indicate Iso-N transition temperature and the solid line the fit to a power-law in temperature.

3.3.4.1 Phase Diagram

The temperature concentration phase diagram, presented in Figure 3.6 (b) depicts the influence of CNT on the host 28% 6OCB in 8OCB binary mixture. Over the entire range of CNT concentration studied, the phase sequence remains the same: isotropic-nematic-SmA-re-entrant nematic (I-N-SmA-RN), as established by deploying several probes. Polarizing optical microscopy exhibits the characteristic fan-shaped and mosaic textures for the smectic and nematic phases, respectively. The attractive feature, however, is the remarkably different influence on the involved phase boundaries. Whereas the clearing point (T_{IN}) diminishes substantially, the transition between the high-temperature N and the smectic phase (T_{N-SmA}) has barely noticeable changes. In contrast, the SmA-RN phase boundary is influenced to the maximum, reducing by more than 11 K, with consequent increase in the thermal range of the SmA by more than 11 K. This stark difference in the behavior of the different boundaries is confounding. The reduction of T_{Iso-N} can be explained from the diminution of the orientational order parameter as suggested by the dielectric anisotropy data. Before discussing the possible causes for the opposite behavior of the two boundaries involving the nematic and smectic phases, let us look at the dielectric, elastic, and structural behavior of the system.

3.3.4.2 Enhanced Antiparallel Ordering

Dielectric results in the mesophases did clearly indicate that the antiparallel ordering of the LC molecules, characteristic of the strongly terminally polar alkyl cyanobiphenyl mesogens, and the associated dimer formation, are enhanced. This favor for antiparallel ordering is evident in the isotropic phase also. The Iso–N transition, being weakly first order, displays pretransitional effects associated with short-range ordering. In liquids of polar materials having dipole moment μ, the permittivity ε_{iso}, being proportional to $\mu^2/k_B T$ (k_B is the Boltzmann constant) increases linearly and monotonically with decreasing temperature. Strongly terminally polar LCs, specifically such as the ones used here, also follow this expectation only deep in the Iso phase: on approaching T_{Iso-N}, the trend reverses after passing through ε_{max}, a maximum in the value,[48–50] a feature referred to as convex-shaped anomaly. Figure 3.6(c) presents such data for the host LC and the LC-CNT composite, both of which display pronounced convex-shaped anomaly feature. It is interesting to note that the coordinates of the peak point, $\delta\varepsilon = (\varepsilon_{max} - \varepsilon_{IN})/\varepsilon_{IN}$ and $\delta T = T_{max} - T_{IN}$, are higher for the composite providing a direct support to the proposal that antiparallel pairing/dimer formation is favored in the presence of CNT. Quantitative description of the data in terms of asymptotic power-law behavior[51] suggests that upon adding CNT, the system becomes closer to a tricritical point.

3.3.4.3 Induction of Layered Phase

Whereas for the concentration (28% 6OCB in 8OCB) of the binary LC system described in the previous section, the layered smectic phase occurs in the phase sequence, but its range gets enhanced upon addition of CNT. For the studies to be discussed in the present section, we chose a host-LC concentration (29.4% 6OCB in 8OCB) lying beyond the apex of the parabola formed by the loci of the N-SmA and SmA-RN phase boundaries. This host mixture (HLC) exhibits only the nematic mesophase. Incorporation of a small quantity of CNT (0.34 wt%) into HLC results in a surprising feature: the SmA phase gets induced by splitting the nematic region into N and RN parts. The induction of the layered phase was evidenced by optical microscopy, laser transmission, elastic constant, and X-ray diffraction probes; in the following paragraph, we will discuss the latter two.

The elastic behavior of the N (and also RN) phase is controlled by three principal Frank elastic constants, referred to as splay (K_{11}), twist (K_{22}), and bend (K_{33}) governing the associated director deformations. In the SmA phase, however, owing to the presence of layering, the twist and bend deformations are not permitted. Hence on approaching the SmA phase from the N (or RN) phase, K_{11} shows a small enhancement, but K_{22} and K_{33} diverge. This feature has attracted much attention since the thermal behavior of K_{33} provides valuable information about the critical phenomenon associated with the one-dimensional positional ordering of the system. A convenient and reliable method for measuring these elastic constants is the magnetic field-driven Freedericksz transformation. Here we describe the results on the thermal behavior of B_{33}, the threshold field for the bend deformation. The samples were sandwiched between two indium-tin-oxide-coated glass plates treated to promote homeotropic alignment of the molecules, and the assembly positioned inside a temperature-controlled stage located between the pole pieces of an electromagnet that provides a maximum field of $B = 2$ T. The magnetic field direction was in the plane of the glass plates, and normal to the initial director orientation. The sample capacitance (C_p) was used as the probe, which yields values corresponding to the equilibrium $C_{\parallel}$ for fields less than a critical value B_{33}, and $C_{\perp}$ for $B \gg B_{33}$. The critical field B_{33} is related to the bend elastic constant K_{33} through

$$B_{33} \propto (\mu_o K_{33}/\chi_a)^{1/2} \tag{3.3}$$

With μ_o as the permeability of free space and χ_a as the diamagnetic susceptibility anisotropy. Figure 3.7(a) presents the thermal behavior of B_{33} for the host LC as well as the HLC+CNT composite. The HLC sample shows a weak thermal dependence. Usually, in the cases where only the nematic phase is present, B_{33} exhibits monotonic variation with temperature without

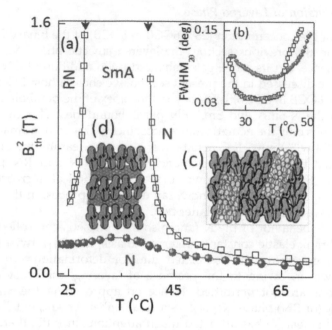

FIGURE 3.7

(a) Temperature dependence of the threshold magnetic field B_{th}^2, for the host LC mixture (filled circles) with 29.4% 6OCB in 8OCB exhibiting only the N phase, and the composite with 0.34% CNT in this mixture (open square), exhibiting the N-SmA-RN sequence. While the no-CNT material has a weak thermal dependence, the CNT composite exhibits critical divergence on approaching the induced SmA phase from both N and RN phases at temperatures indicated by the downward arrows. (b) The thermal variation of the full-width-at-half-maximum of the X-ray diffraction profiles at low angles for the two materials. The material with CNT exhibits a sharp drop in the value at temperatures corresponding to the transition to the SmA phase. Two possible dispositions of CNT in the LC environment (shown as dimers with the arrow heads showing the CN dipole in the molecule): the long axis of CNT is parallel to the director in (c) and orthogonal to it in (d). The latter represents the nanophase segregation situation, which we argue could be a possible cause for induction of SmA phase.

a maximum. Thus, the weak peak observed could be due to the system being close to the inflection point (the tip of the parabola) of the binary phase diagram. In contrast, the CNT composite exhibits strong diverging behavior on approaching ~45 °C and 28 °C, from above and below, respectively. Owing to the larger temperature range available for analysis, we consider the data on the N-SmA side only. The theory[52] expects the temperature dependence of K_{33}, or equivalently B_{33}, to be given by the power-law,

$$B_{33}^2 = B_1 \left(\frac{T - T_{AN}}{T_{AN}} \right)^{-x} + B_2 \tag{3.4}$$

where B_1 and B_2 are constants, and T_{AN}, the N-SmA transition tempera-ture, and x is the same exponent that governs the temperature-dependent smectic correlation length. Equation (3.4) describes the data very well as seen in Figure 3.7(a) yielding the exponent $x = 0.65 \pm 0.02$, a value quite close to that expected for the XY universality class ($x \sim 0.66$) to which this transition belongs to.

We now discuss the structural evidence to establish that the induced phase is indeed a smectic. In fact, the difference between the N phase and the (induced) smectic can be made out even at a qualitative level, by visual means of looking at the diffraction profiles at low angles. For HLC, over the entire temperature range of the mesophase, the profile remains, as expected, broad and diffuse, but for a minimum in the profile width (full-width-at-half-maximum) at temperatures corresponding to the tip of parabolic phase boundary. The behavior of the CNT composite is distinctly different: in the high- and low-temperature regions, the profile is broad and diffuse, but sharp and intense in the intermediate region corresponding to the induced SmA phase. The variation of the profile width as a function of temperature for the two materials, shown in Figure 3.7(b), brings out these features. Having provided unambiguous evidence for the enhancement of the layered structure as well as its induction, we now return to the possible cause for the suppression of RN and concomitant stabilization of SmA in the 28% 6OCB host mixture, and induction of the smectic phase in the 29.4% 6OCB mixture. Induction of the smectic phase in an otherwise-nematic-only mate-rial has been ascribed to features such as hydrogen bonding, charge transfer complex formation, dipole–dipole, dipole-induced dipole interactions,[53] or photo-driven nanosegregation.[54] For the system under consideration since parallels can be drawn to the photo-driven dynamic self-assembly, we present it as one possibility for the induction/enhancement of the layered phase. In the photo-driven case, some of the constituents exhibit the phe-nomenon of trans-cis photoisomerization, wherein the photo-isomerized cis state, when present, induces the SmA, or if already existing, enhances its thermal range, just as CNTs do in the present case. An explanation for the photo-driven stabilization of SmA has been provided using the photo segregation model. In this model, the shape-altered (from rod-like to bent) photoactive molecules promote a local-level demixing from the rod-like host molecules, creating conditions favorable for the appearance/stabilization of the layered phase. Applying this idea to the present case, it is proposed that while the system favors N (or RN) phase with the CNTs being parallel to the director (see Figure 3.7(c)), their inherent tendency would be to create nanophase segregation wherein the local-level demixing of CNT and LC places the CNT such that layering of the host molecules is stabilized (Figure 3.7(d)). A second possibility for the observations described is the frustration caused in the packing of the LC molecules when CNTs are present. Being on the lines of the "Frustrated Spin-gas" model,[55] specifically applicable to molecules with a strongly polar terminal group, this is well suited for the present case. The model perceives intermolecular steric hindrance and van der Waals attraction but focuses on the dipole–dipole forces among a triplet of polar host molecules. Balancing the dipolar contribution to the free energy from such a triplet and the ability of the molecules to liberate along the director direction, either the N (or RN) and the SmA can be stabilized. As per this theory, short-range dipolar interactions dominate and the triplets favor layering. Both the nanophase segregation and the frustrated spin-gas situations have more effect on the RN side of the phase sequence owing to thermal fluctuations. For the high temperature N-SmA boundary thermal fluctuations dominate, thus reducing the effect due to the other causes mentioned above. This explains strikingly larger effects of CNT on the SmA-RN boundary than on the N-SmA boundary. Raman spectroscopy measurements would lend support to some of these ideas.

References

1. Li, Q. *Nanoscience with Liquid Crystals: From Self-Organized Nanostructures to Applications*. Springer Cham Heidelberg; **2013**.
2. Hegmann, T.; Qi, H.; Marx, V. M. Nanoparticles in Liquid Crystals: Synthesis, Self Assembly, Defect Formation and Potential Applications. *J. Inorg. Organomet. Polym. Mater.* **2007**, *17*, 483–508.
3. Lagerwall, J. P. F.; Scalia, G. *Liquid Crystals with Nano and Microparticles*. World Scientific Publishing Co. Pvt. Ltd, Singapore; **2017**.
4. Yadav, S. P.; Singh, S. Carbon Nanotube Dispersion in Nematic Liquid Crystals: An Overview. *Prog. Mater. Sci.* **2016**, *80*, 38–76.
5. Dierking, I.; Scalia, G.; Morales, P.; LeClere, D. Aligning and Reorienting Carbon Nanotubes with Nematic Liquid Crystals. *Adv. Mater.* **2004**, *16*, 865–869.
6. See for e.g., Chandrasekhar, S. *Liquid Crystals*. Cambridge University Press, Cambridge; **1992**.
7. Wu, S.-T.; Yang, D.-K. *Fundamentals of Liquid Crystal Devices (Wiley Series in Display Technology)*, John Wiley & Sons, Ltd, Chichester; **2015**.
8. Jayalakshmi, V.; Prasad, S. K. Understanding the Observation of Large Electrical Conductivity in Liquid Crystal-Carbon Nanotube Composites. *Appl. Phys. Lett.* **2009**, *94*, 202106.
9. Koshio, A.; Yudasaka, M.; Zhang, M.; Iijima, S. A Simple Way to Chemically React Single-Wall Carbon Nanotubes with Organic Materials Using Ultrasonication. *Nano Lett.* **2001**, *1*, 7, 361–363.
10. Prakash, J.; Choudhary, A.; Mehta, D. S.; Biradar, A. M. Effect of Carbon Nanotubes on Response Time of Ferroelectric Liquid Crystals. *Phys. Rev. E* **2009**, *80*, 012701.
11. Lagerwall, J. P. F.; Dabrowski, R.; Scalia, G. Antiferroelectric Liquid Crystals with Induced Intermediate Polar Phases and the Effects of Doping with Carbon Nanotubes. *J. Non. Cryst. Solids* **2007**, *353*, 4411–4417.
12. Krishna Prasad, S.; Vijay Kumar, M.; Yelamaggad, C. V. Dual Frequency Conductivity Switching in a Carbon Nanotube/Liquid Crystal Composite. *Carbon N. Y.* **2013**, *59*, 512–517.

13. Lebovka, N.; Goncharuk, A.; Bezrodna, T.; Chashechnikova, I.; Nesprava, V. Microstructure and Electrical Conductivity of Hybrid Liquid Crystalline Composites Including 5CB, Carbon Nanotubes and Clay Platelets. *Liq. Cryst.* **2012**, *39*, 531–538.

14. Krishna Prasad, S.; Sandhya, K. L.; Nair, G. G.; Hiremath, U. S.; Yelamaggad, C. V.; Sampath, S. Electrical Conductivity and Dielectric Constant Measurements of Liquid Crystal–Gold Nanoparticle Composites. *Liq. Cryst.* **2006**, *33*, 1121–1125.

15. Prasad, S. K.; Kumar, M. V.; Shilpa, T.; Yelamaggad, C. V. Enhancement of Electrical Conductivity, Dielectric Anisotropy and Director Relaxation Frequency in Composites of Gold Nanoparticle and a Weakly Polar Nematic Liquid Crystal. *RSC Adv* **2014**, *4*, 4453–4462.

16. Sridevi, S.; Prasad, S. K.; Nair, G. G.; D'Britto, V.; Prasad, B. L. V. Enhancement of Anisotropic Conductivity, Elastic, and Dielectric Constants in a Liquid Crystal-Gold Nanorod System. *Appl. Phys. Lett.* **2010**, *97*, 151913.

17. Lakshmi Madhuri, P.; Krishna Prasad, S.; Shinde, P.; Prasad, B. L. V. Large Reduction in the Magnitude and Thermal Variation of Frank Elastic Constants in a Gold Nanorod/Nematic Composite. *J. Phys. D. Appl. Phys.* **2016**, *49*, 425304.

18. Kamaliya, B.; Vijay Kumar, M.; Yelamaggad, C. V.; Krishna Prasad, S. Enhancement of Electrical Conductivity of a Liquid Crystal-Gold Nanoparticle Composite by a Gel Network of Aerosil Particles. *Appl. Phys. Lett.* **2015**, *106*, 083110.

19. Lebovka, N.; Dadakova, T.; Lysetskiy, L.; Melezhyk, O.; Puchkovska, G.; Gavrilko, T. Phase Transitions, Intermolecular Interactions and Electrical Conductivity Behavior in Carbon Multiwalled Nanotubes/Nematic Liquid Crystal Composites. *J. Mol. Struct.* **2008**, *887*, 135–143.

20. Jonscher, A. K. Frequency-Dependence of Conductivity in Hopping Systems. *J. Non. Cryst. Solids* **1972**, *8–10*, 293–315.

21. Cramer, C.; Brunklaus, S.; Ratai, E.; Gao, Y. New Mixed Alkali Effect in the AC Conductivity of Ion-Conducting Glasses. *Phys. Rev. Lett.* **2003**, *91*, 266601.

22. Louati, B.; Gargouri, M.; Guidara, K.; Mhiri, T. AC Electrical Properties of the Mixed Crystal $(NH_4)3H(SO_4)1.42(SeO_4)0.58$. *J. Phys. Chem. Solids* **2005**, *66*, 762–765.

23. Papathanassiou, A. N.; Sakellis, I.; Grammatikakis, J. Universal Frequency-Dependent Ac Conductivity of Conducting Polymer Networks. *Appl. Phys. Lett.* **2007**, *91*, 122911.

24. De Jeu, W. J.; Gerritsma, C. J.; Van Zanten, P.; Goossens, W. J. A. Relaxation of the Dielectric Constant and Electrohydro-Dynamic Instabilities in a Liquid Crystal. *Phys. Lett. A* **1972**, *39*, 355–356.

25. *Smart Glass and Windows 2018–2028: Electronic Shading and Semi-Transparent PV.* IDTechEx. www.idtechex.com/research/reports/smart-glass-and-windows-2018-2028-electronic-shading-and-semi-transparent-pv-000601.asp (accessed Feb 7, 2019).

26. Hinojosa, A.; Sharma, S. C. Effects of Gold Nanoparticles on Electro-Optical Properties of a Polymer-Dispersed Liquid Crystal. *Appl. Phys. Lett.* **2010**, *97*, 081114-1-3.

27. Garbovskiy, Y.; Glushchenko, A. Ferroelectric Nanoparticles in Liquid Crystals: Recent Progress and Current Challenges. *Nanomaterials* **2017**, *7*, 361.

28. Rahman, M.; Lee, W. Scientific Duo of Carbon Nanotubes and Nematic Liquid Crystals. *J. Phys. D. Appl. Phys* **2009**, *42*, 063001-1-12.

29. Basu, R.; Iannacchione, G. S. Dielectric Hysteresis, Relaxation Dynamics, and Nonvolatile Memory Effect in Carbon Nanotube Dispersed Liquid Crystal. *J. Appl. Phys* **2009**, *106*, 124312–124312-6.

30. Kim, E.; Liu, Y.; Hong, S.-J.; Han, J. I. Effect of SiO2 Nanoparticle Doping on Electro-Optical Properties of Polymer Dispersed Liquid Crystal Lens for Smart Electronic Glasses. *Nano Converg* **2015**, *10*, 607–610.

31. Abbasov, M. E.; Carlisle, G. O. Effects of Carbon Nanotubes on Electro-Optical Properties of Dye-Doped Nematic Liquid Crystal. *J. Mater. Sci. Mater. Electron.* **2012**, *23*, 712–717.

32. Deshmukh, R. R. *Liquid Crystalline Polymers, Vol.2-Processing and Applications*. V. K. Thakur, & M. R. Kessler, Eds. Springer International Publishing: Switzerland, **2015**.

33. Liu, B.; Ma, Y.; Zhao, D.; Xu, L.; Liu, F.; Zhou, W.; Guo, L. Effects of Morphology and Concentration of CuS Nanoparticles on Alignment and Electro-Optic Properties of Nematic Liquid Crystal. *Nano Res* **2016**, *10*, 618–625.

34. Krishna Prasad, S.; Baral, M.; Murali, A.; Jaisankar, S. N. Carbon Nanotube Reinforced Polymer-Stabilized Liquid Crystal Device: Lowered and Thermally Invariant Threshold with Accelerated Dynamics. *ACS Appl. Mater. Interfaces* **2017**, *9*, 26622–26629.

35. Sapsanis, C.; Omran, H.; Chernikova, V.; Shekhah, O.; Belmabkhout, Y.; Buttner, U.; Eddaoudi, M.; Salama, K. N. Insights on Capacitive Interdigitated Electrodes Coated with MOF Thin Films: Humidity and VOCs Sensing as a Case Study. *Sensors* **2015**, *15*, 18153–18166.

36. Narayanan, T.; Kumar, A. Reentrant Phase Transitions in Multicomponent Liquid Mixtures. *Phys. Rep.* **1994**, *249*, 135–218.

37. For an early review see, Cladis, P. E. A One Hundred Year Perspective of the Reentrant Nematic Phase. *Mol. Cryst. Liq. Cryst. Inc. Nonlinear Opt.* **1988**, *165*, 85–121.

38. Litster, J. D.; Birgeneau, R. J. Phases and Phase Transitions. *Phys. Today* **1982**, *35*, 26–33.

39. Kolafa, J.; Nezbeda, I.; Pavlíček, J.; Smith, W. R. Global Phase Diagrams of Model and Real Binary Fluid Mixtures: Lorentz–Berthelot Mixture of Attractive Hard Spheres. *Fluid Phase Equilibria* **1998**, *146*, 103–121.

40. Kim, Y.; Alexe, M.; Salje, E. K. H. Nanoscale Properties of Thin Twin Walls and Surface Layers in Piezoelectric WO3–x. *Appl. Phys. Lett.* **2010**, *96*, 032904.

41. Strečka, J.; Ekiz, C. Reentrant Phase Transitions and Multicompensation Points in the Mixed-Spin Ising Ferrimagnet on a Decorated Bethe Lattice. *Phys. A Stat. Mech. Its Appl.* **2012**, *391*, 4763–4773.

42. Heera, V.; Fiedler, J.; Hübner, R.; Schmidt, B.; Voelskow, M.; Skorupa, W.; Skrotzki, R.; Herrmannsdörfer, T.; Wosnitza, J.; Helm, M. Silicon Films with Gallium-Rich Nanoinclusions: From Superconductor to Insulator. *New J. Phys.* **2013**, *15*, 083022.

43. Möller, J.; Grobelny, S.; Schulze, J.; Bieder, S.; Steffen, A.; Erlkamp, M.; Paulus, M.; Tolan, M.; Winter, R. Reentrant Liquid-Liquid Phase Separation in Protein Solutions at Elevated Hydrostatic Pressures. *Phys. Rev. Lett.* **2014**, *112*, 028101.

44. Dehyadegari, A.; Sheykhi, A. Reentrant Phase Transition of Born-Infeld-AdS Black Holes. *Phys. Rev. D* **2018**, *98*, 024011.

45. Yelamaggad, C. V.; Shashikala, I. S.; Shankar Rao, D. S.; Nair, G. G.; Krishna Prasad, S. The Biaxial Smectic (SmAb) Phase in Nonsymmetric Liquid Crystal

Dimers Comprising Two Rodlike Anisometric Segments: An Unusual Behavior. *J. Mater. Chem.* **2006**, *16*, 4099.

46. Pociecha, D.; Gorecka, E.; Čepič, M.; Vaupotič, N.; Žekš, B.; Kardas, D.; Mieczkowski, J. Reentrant Ferroelectricity in Liquid Crystals. *Phys. Rev. Lett.* **2001**, *86*, 3048–3051.

47. Guillon, D.; Cladis, P. E.; Stamatoff, J. X-Ray Study and Microscopic Study of the Reentrant Nematic Phase. *Phys. Rev. Lett.* **1978**, *41*, 1598.

48. Bradshaw, M. J.; Raynes, E. P. Pre-Transitional Effects in the Electric Permittivity of Cyano Nematics. *Mol. Cryst. Liq. Cryst.* **1981**, *72*, 73–78.

49. Janik, M.; Rzoska, S. J.; Rzoska, A. D.; Zioło, J.; Janik, P.; Maslanka, S.; Czupryński, K. Pretransitional Behavior in the Isotropic Phase of a Nematic Liquid Crystal with the Transverse Permanent Dipole Moment. *J. Chem. Phys.* **2006**, *124*, 144907.

50. Sridevi, S.; Prasad, S. K.; Rao, D. S. S.; Yelamaggad, C. V. Pretransitional Behaviour in the Vicinity of the Isotropic–Nematic Transition of Strongly Polar Compounds. *J. Phys: Condens Matter* **2008**, *20*, 465106.

51. Drozd-Rzoska, A.; Rzoska, S. J.; Czuprynski, K. Phase Transitions from the Isotropic Liquid to Liquid Crystalline Mesophases Studied by Linear and Nonlinear Static Dielectric Permittivity. *Phys. Rev. E. Stat. Phys. Plasmas. Fluids. Relat. Interdiscip. Topics* **2000**, *61*, 5355–5360.

52. Jähnig, F.; Brochard, F. Critical Elastic Constants and Viscosities above a Nematic-smectic a Transition of Second Order. *Journal de Physique* **1974**, *35*, 301–313.

53. Sugisawa, S.-Y.; Tabe, Y. Induced Smectic Phases of Stoichiometric Liquid Crystal Mixtures. *Soft Matter* **2016**, *12*, 3103.

54. Prasad, S. K.; Nair, G. G.; Hegde, G. Dynamic Self-Assembly of the Liquid-Crystalline Smectic A Phase. *Adv. Mater.* **2005**, *17*, 2086–2091.

55. Indekeu, J. O.; Berker, A. N.; Nihat Berker, A. Molecular Structure and Reentrant Phases in Polar Liquid Crystals. *J.Phys. France* **1988**, *49*, 353–362.

4

Selective Segregation and Crystallization Induced Organization of Carbon Nanotube Network in Polymer Nanocomposites

Aishwarya V. Menon

Center for Nano Science and Engineering
Indian Institute of Science
Bangalore, India

Tanyaradzwa S. Muzata and Suryasarathi Bose

Department of Materials Engineering,
Indian Institute of Science
Bangalore, India

4.1 Introduction

Carbon nanotubes (CNTs) have attracted tremendous attention in recent times owing to their superb electronic, optical, and mechanical properties. From the consistent research efforts, it has been proven that CNTs are not only useful and interesting systems in themselves but can also show interesting properties when used to fabricate nanocomposites. CNTs are often used to reinforce polymeric matrices for the electronic industry to dissipate the unwanted static charge build-up and in automotive applications to facilitate electrostatic painting. The high aspect ratio of CNTs allows them to impart polymer/CNT composites with high stiffness and good electrical conductivity at relatively low loading of CNT.[1]

At present, numerous approaches such as solution mixing using a suitable solvent and melt blending have been successfully established to incorporate CNTs in polymer matrices.[2,3] Despite these approaches, it is still a challenge to practically fabricate CNT/polymer composites with physical properties that approach the maximum theoretical values of individual CNTs. Numerous significant advances have been made to date to fill the gap between expectations and practical performance. Directional alignment of CNTs in the polymer matrix is quite helpful in this context, which helps to achieve properties that are close to theoretical values.

Despite the versatile properties of aligned CNT/polymer composites, their preparation with a simple and straightforward solution or melt blending approach may compromise the dispersion and alignment of CNTs. Even under experimental conditions, a well-aligned assembly of CNTs tends to become isotropic and clustered when incorporated into the composite, eventually leading to random orientation.[4,5] Therefore, feasible approaches toward alignment that can be applicable to a wide range of polymer matrices are highly desirable[6]. CNTs can be aligned vertically or horizontally in the polymer matrix. The horizontal or vertical alignment can be achieved by the electric field and magnetic field; mechanical stretching; electrospinning, using shear forces; infiltration and so on. There are numerous literature available that discuss various techniques used till date to preferentially align CNTs.[7–9]

However, the scope of this chapter lies in discussing the various methods that can be used to form organized networks of CNTs in polymer nanocomposites utilizing the various strategies such as crystallization-driven, phase separation, order–disorder transition, crosslinking/curing, or a combination of these. These strategies besides facilitating an organized network of CNTs also result in intriguing properties such as low percolation threshold in binary blends, enhanced rates of crystallization, and influencing the degree of crystallinity. Such a comprehensive review will certainly guide researchers working in this area from both academia and industry.

4.2 Organization of CNTs during Polymer Crystallization

4.2.1 Nucleating Agent and Polymorphism

From the existing literature, CNTs have been found to act as nucleating agents for polymer crystallization. They also act as templates for polymer chain orientation.[10] The high aspect ratio coupled with their substantial surface area make CNTs, especially single-walled CNTs (SWCNTs), stand out from other already known nucleating agents.[11] Bhattacharya et al.[12] reported an increase in the isothermal crystallization rate, narrower crystallization, melting peaks, and smaller spherulite size in PP/SWCNT composites. The changes observed in the crystallization behavior and spherulitic morphology were attributed to the enhanced nucleation of PP in the presence of SWCNT. In a similar study by Valentini et al.,[13] it was reported that the presence of SWCNTs enhanced the nucleation process and this process showed saturation at very low concentration. The crystallization kinetics was found to be strongly affected by the inter-nanotube bundle distance. The reduction in the degree of crystallinity of PP was confirmed from the decrease in the heat of fusion. It was also found that the presence of nanotubes decreased the size of spherulites from 100 μm for neat PP to 10 μm for PP/SWCNT composites. Grady et al.[14] prepared PP/SWCNT nanocomposites by solution blending method. It was found that the presence of SWCNTs promoted the growth of the less preferred β phase of crystalline PP. The crystallization rate and the percent crystallinity were also found to be higher in nanocomposites, thus confirming the enhancement in nucleation of PPP.[12,14–17] The nucleating effect of CNT besides PP has also been observed for many other polymers including polyamide (Nylon 6 and Nylon 6,6), polyethylene terephthalate, poly (1-butene), poly(vinyl alcohol), polyethylene (PE) and so on.[18–27] Because of the nucleation effect of CNT, crystallization temperature gets shifted to a higher temperature and crystallization rate becomes much faster in the nanocomposites compared to virgin polymer.

4.2.2 Formation of Shish–Kebab Morphology

The high aspect ratio, cylindrical shape, nanometer size diameter, and graphitic structure of CNTs make it possible for polymer chains to align along their axis. Several polymers such as PE and Nylon-6,6 form shish–kebab-like morphology along the CNT axis when crystallized by solution crystallization as shown in Figure 4.1.[28] In these cases, CNT serves as the shish and causes the orientation of growth direction of polymer lamellas perpendicular to its axis. These nanohybrid shish–kebab (NHSK) structures are emerging as one of the noncovalent functionalization techniques to achieve functionalized CNTs. The NHSK structures have gained massive attention due to a variety of factors, which include easy side or end

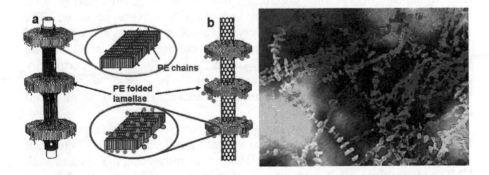

FIGURE 4.1
Schematic representations of (a) shish–kebab crystals in pristine PE and (b) PE/CNT NHSK structure.

functionalization of CNTs; noncovalent nature of the technique, which does not disturb inherent CNT structure; easily controllable degree of functionalization by changing the kebab size, polymer solution concentration, and crystallization time; periodic nature of the functionalization, which provides a unique approach to create functional and ordered structure along CNTs for electrical and optical applications.[29-32] A classic shish–kebab crystal consists of a central fibril (known as the "shish") and multiple disc-shaped folded-chain lamellae (known as the "kebabs") orthogonally placed with respect to the shish. In case of pristine polymer, both the shish and kebab are formed by the polymer. In case of polymer/CNT nanocomposites, the shish is made of CNTs instead of the polymer crystals, hence the name "nanohybrid shish–kebab" was coined.

Obviously, the mechanism of formation of NHSK and the classical shish–kebabs is different as can be visualized from Figure 4.1. In case of classic shish–kebab, polymer solutions need to be placed under an extensional flow. The polymer chains that normally possess a coil conformation might be stretched and if the chain is longer than a critical molecular weight, it aggregates to form extended chain fibrillar crystals. The remaining coiled polymer chains would then crystallize upon the fibrillar crystals in a folded, periodic fashion, forming the shish–kebab morphology due to linear nucleation.[33,34] The various mechanisms of NHSK formation are discussed below.

CNTs are unique compared to other nanofillers because of two distinct properties: (1) their aspect ratio and size and (2) their surface chemistry. Solution crystallization has emerged as a novel process to produce ideal polymer single crystals. Li et al. studied two model polymers, PE and Nylon 6,6, to observe the NHSK structure. p-Xylene and 1,2-dichlorobenzene (DCB) were used as the solvents for controlled solution crystallization in the case of PE.[28,31, 35,36] The heterogeneous nucleation of PE was facilitated by keeping the temperature of crystallization higher than the clearing temperature of PE in p-xylene. Disc-shaped decorations on the CNTs were observed. The decorations were edge-on PE single-crystal lamellae with an average lateral dimension of 50–80 nm. It was interesting to note that the PE lamellae were held together by multiwalled CNTs (MWCNTs) with the average periodicity of ~40 to 50 nm. The morphology was similar to the classic "shish–kebab" polymer crystals that were formed under an extensional field, as observed by Geil and Pennings in 1960s.[37,38]

In the study by Li et al., the nano-fibrillar structure of CNTs provided a 1D nucleation surface; therefore, shear was not required in the CNT-induced PE crystallization to form NHSK. CNTs serve as nucleating agents and each CNT had multiple nucleation sites. Not only MWCNTs, but SWCNT was also found to form NHSK with PE. The orthogonal orientation between PE lamellar surface and the CNT axis indicated that the PE chains were parallel to the CNT axis. They established two possible factors that could be the reason behind NHSK growth: first being the epitaxial growth of PE on CNT and second being the geometric confinement. The first factor was attributed to the well-established epitaxial growth of PE onto highly ordered pyrolytic graphite.[39,40] The difference in chirality of the CNTs gives rise to multiple orientations of its graphitic lattice with respect to the CNT axis. However, if the epitaxy was the only dominating factor, multiple orientations of the PE single-crystal lamellae must be observed, which was contradictory to what was observed. This is where the second factor comes into play. Since the diameter of CNT is quite similar to the radius of gyration of polymer chains, its highly curved surface causes strong geometric confinement forcing the polymer chains to align parallel to the CNT axis upon crystallization. This occurs regardless of the CNT chirality. Although a strict crystal lattice matching between CNT surface and polymer crystals is not required, the cooperative orientation of the polymer chains and the CNT axes is a must. The term "soft epitaxy" was coined to describe such unique growth mechanism as illustrated in Figure 4.2.

Similar NHSK structures were reported by several groups. Zhang et al.[41] sprayed water soluble, sodium dodecyl sulfate coated SWCNTs on the surface

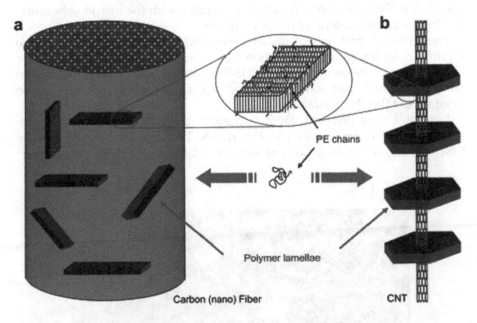

FIGURE 4.2
Schematic representation of the "size-dependent soft epitaxy" mechanism. (a) For large-diameter CNFs, PE lamellae are randomly orientated on the fiber surface. (b) For small-diameter CNTs, soft epitaxy dictates the parallel orientation between PE chains and the CNT axis, leading to an orthogonal orientation between CNT and PE lamellae.

Reprinted from Li et al.[28] Copyright 2006 with permission from American Chemical Society.

of ultrahigh molecular weight PE (UHMWPE). The mixture was then dissolved in xylene at a relatively high concentration. The mixture was allowed to cool and crystallize for a week forming a gel. Similarly, Uehaha et al.[42] used DCB to induce solution crystallization of UHMWPE. In an interesting study by Zhang et al.,[43] supercritical carbon dioxide was used as a nonsolvent to induce PE crystallization from PE/xylene or PE/DCB solution.

Similarly, studies on CNT-induced formation of NHSK on Nylon 6,6 revealed that, upon crystallization, Nylon 6,6 adopts a planar zigzag conformation, which facilitates CNT-induced crystallization as shown in Figure 4.4. Furthermore, it was also been demonstrated that Nylon can epitaxially grow on any highly ordered pyrolytic graphite surface.[44] Li et al.[45] modified MWCNTs with Nylon 6,6 via a controlled polymer solution crystallization method. An NHSK structure was achieved, wherein the MWCNT resembled the shish while Nylon 66 lamellar crystals formed the kebabs as shown in Figure 4.3.

These Nylon 6,6-functionalized MWCNTs were then used as precursors to prepare polymer/MWCNT nanocomposites, whereby a Nylon 66–glycerol solution was added to the precursor NHSK suspension at T_c (185 °C) and allowed to further crystallize for 3 h. The SEM micrographs in Figure 4.4 shows that Nylon 66 forms rounded spherulites with the hybrid spherulites having an average diameter of about 10 μm.

The formed hybrid nanocomposites were etched with nitric acid and from Figure 4.5 an intriguing observation was revealed. MWCNT networks are clearly seen in the etched areas. NHSK structures were occasionally seen from the etched area indicating good interfacial adhesion of Nylon 66 crystals to MWCNT surface.

Li et al.[46] used physical vapor deposition to study CNT-induced polymer crystallization as shown in Figure 4.6. They placed a small drop of SWCNT dispersion in a solvent such as DCB on a carbon-coated cover

FIGURE 4.3
(a) SEM image of Nylon 66-functionalized MWCNT; (b) TEM image of the similar structure.

Reprinted from Li et al.[45] Copyright 2007 with permission from Elsevier.

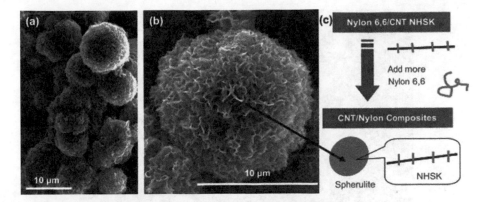

FIGURE 4.4
(a) and (b) are the SEM micrographs of 0.5 wt% CNT/Nylon 66 nanocomposites formed by using NHSK as precursor.(c) Schematic of the nanocomposite preparation.

Reprinted from Li et al.[45] Copyright 2007 with permission from Elsevier.

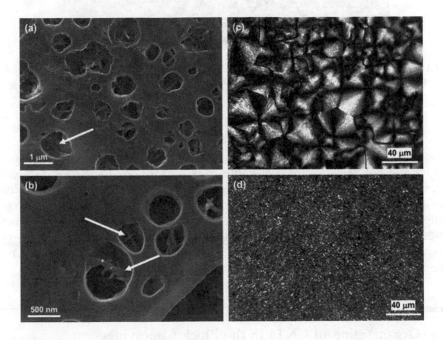

FIGURE 4.5
(a) SEM micrograph of Nylon 66/MWCNT nanocomposites after nitric acid etching.(b) Higher magnification SEM micrograph of NHSK found in the etched areas.(c) and (d) are optical microscope images of Nylon 66 and Nylon 66/MWCNT nanocomposite films. The films were prepared by the melt-mixing.

Reprinted from Li et al.[45] Copyright 2007 with permission from Elsevier.

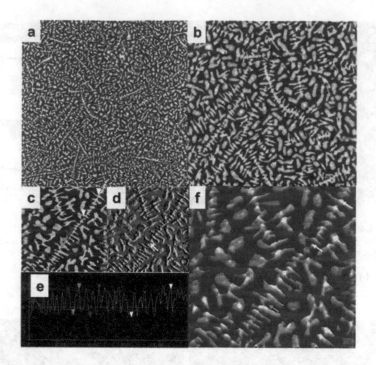

FIGURE 4.6
AFM Tapping-Mode Images of the PE-Decorated SWCNTs.

Reprinted from Li et al.[46] Copyright 2006 with permission from American Chemical Society.

glass or a TEM grid using a spin coater, after which it was exposed to the vaporized polymer by physical vapor deposition. They used four different types of polymers: PE, Nylon 6,6, polyvinylidene fluoride, and poly-L-lysine for this study. AFM image revealed many small "islands" with an average height of ~10 nm. It was seen that as compared to the rest of the polymer "islands" formed on the carbon film, the polymer rods attached to SWCNTs showed uniform orientation whereby they were semiperiodically located on the SWCNTs.

4.3 Organization of CNTs in the Block Copolymer

Nanoelectronics application calls for the creation of ordered arrays of CNTs for large-scale integrated circuits, which is an area where significant progress has been made, and for the production of controllable patterns on individual CNTs to fabricate multiple transistors on them, which is an area where progress has been relatively slower.[47–52] In this context, judiciously selected crystalline block copolymers can be useful, whereby they can be periodically decorated along with CNTs, leading to the formation of amphiphilic alternating patterns, with a period of ~12 nm. Till date, there are very few reports which have addressed periodic functionalization or patterning on CNTs.[32,52–55] Also, most of the reported periodic patterns to date suffer from limitations such as poor periodicity or locality of the pattern.

Li et al.[56] reported the use of polyethylene-poly (ethylene oxide) (PE-b-PEO) block copolymer with a low molecular weight to obtain uniform and periodic patterns on CNTs as shown in Figure 4.7. They drop-casted SWCNT/DCB solution on a carbon-coated nickel grid and dried it at ambient temperature. The fractionated PE-b-PEO (having molecular weight ~ 1700 g mol^{-1} and ~50 wt% PE content) was dissolved in chloroform and then spin-coated onto the as-prepared grid. To enhance the contrast, the sample was stained with ruthenium tetroxide (RuO$_4$). They observed the formation of nanoscale alternating patterns of block copolymer along SWCNTs. The period of the patterns was found to be ~12 nm along the SWCNT axis. They attributed the pattern formation to the interplay of CNT-induced PE crystallization and the block copolymer phase separation.

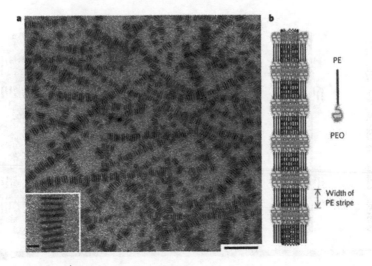

FIGURE 4.7
(a) TEM micrograph of PE-b-PEO decorated SWCNTs. The dark and bright stripes are the PEO and PE domains, respectively;(b) schematic representation of the arrangement of the PE-b-PEO molecules along SWCNT.

They also proposed a growth mechanism for the formation of patterns of alternating stripes on SWCNTs as depicted by Figure 4.8. According to them, the block copolymer molecules randomly adsorb onto the SWCNT surface during spin coating, due to the favorable interaction between PE segments and SWCNTs, causing heterogeneous nucleation. Once a stable nucleus was formed, the PE crystal started to grow, according to the soft epitaxy mechanism previously reported by them.[35] The block copolymer molecules crystallize only to the initially formed nuclei at an extremely low block copolymer concentration. The local concentration gradient generated at the crystal growth front, causes the stripes to grow laterally on the carbon film in an orientation that is perpendicular to the tube axis. CNT-induced PE crystallization essentially drives this process. More crystals start appearing as the block copolymer concentration increases. The PEO chains tethered at the crystal edges and dangling in the solution attract PEO segments of the remaining free block copolymer molecules to the adjacent region and facilitate crystallization at a certain distance away from the initially formed nuclei on the SWCNT. This distance depends on the length of the PEO block and the process is governed by the phase separation of the block copolymer. When the above process is repeated, patches of alternating block copolymer stripes along the SWCNT are formed. Since PE is hydrophobic and PEO is hydrophilic, this particular system represents a hybrid structure on individual CNTs having alternating 12-nm period amphiphilicity and

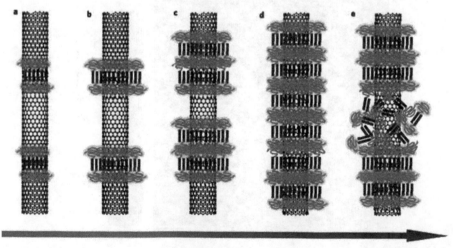

Concentration increasing

FIGURE 4.8

(a) Heterogeneous nucleation of the PE segments on SWCNT. (b–e) Growth patterns of the stripes at various block copolymer concentrations.

Reprinted from Li et al.[56] Copyright 2009 with permission from Springer Nature.

also the size of the bright stripes along the CNT axes (domain size) is 5.9 ± 0.7 nm.[56-58] The length of the period depends on the type of block polymer selected. Since, in the above work, PEO is hydrophilic and PE is hydrophobic, the system exhibited an alternating amphiphilicity with around 12-nm period. Moreover, as the block copolymer molecular weight decreases, the domain size also decreases.

Using a similar approach, Yu et al.[59] successfully modified PE-b-PEO on SWCNTs and MWCNTs using a simple supercritical carbon dioxide antisolvent-induced polymer epitaxy method. They found that, when DCB or *p*-xylene was used as the solvent, the periodic patterns were formed by PE-b-PEO on CNTs; however, when the solvent was switched to *N,N*-dimethylacetamide, which was more selective toward PEO, periodic patterns were not observed. Similarly, Park et al.[60] synthesized nanohybrids consisting of SWCNTs and perchlorate-doped poly(3,4-ethylenedioxythiophene)-block-poly (ethylene oxide) (P-PEDOT-b-PEO), which is a conductive block copolymer. Wang et al.[61] used both PE-b-PEO and PE-b-SBR to form regular and periodic patterns on SWCNTs with a periodicity of ~12–30 nm along the SWCNT axis for the patterns.

Chen et al.[62] developed thermoplastic polyurethane/SWCNT having exceptional alignment and improved mechanical properties using simple compounding technique. The alignment of SWCNTs was achieved by dissolving SWCNTs in a solution of thermoplastic polyurethane and tetrahydrofuran. The driving force for the macroscopic alignment was attributed to solvent–polymer interactions, which induce the orientation of soft chain segments during the swelling and moisture curing stage.

4.4 The Key Role of Segregated Network Formation in CNT Organization

The percolation threshold in polymer nanocomposites can be lowered using fibrous/high aspect ratio fillers; however, their mechanical properties become increasingly dependent on the interfacial wetting, distribution, and orientation of the filler, and the processing technique becomes increasingly complex. Formation of segregated networks in polymer nanocomposites has been a very popular technique to achieve high interparticle connectivity at relatively low nanoparticle loadings.[63-66] Mierczynska et al.[67] demonstrated the formation of the segregated network in PE/MWCNT composites using three different qualities of CNTs by dry blending in a mortar and then sintering in a mold at 5 MPa pressure at a temperature above the melting point of PE. The commercial grade SWCNTs with diameter of 12–15 Å and purity of 50–70% was purchased from Aldrich. The other two types of SWCNTs were produced by arc method in He atmosphere using a Ni/Y catalyst. The high-quality grade was obtained from the surroundings of the cathode while the low-quality grade was obtained from the chimney. The high-quality grade consisted majorly of SWCNTs while the low-quality grade contained significant amounts of amorphous carbon. They achieved electrical percolation threshold of 0.5, 1, and 1.5 wt% for high-quality grade, commercial grade, and low-quality grade, respectively. Lisunova et al.[68] prepared UHMWPE/MWCNT composites by mechanical mixing. The powder was then homogenized in a porcelain mortar to a visually homogeneous state. They achieved low electrical percolation thresholds due to the high aspect ratio of the MWCNTs and their segregated distribution inside the polymer matrix.

Du et al.[69] comparedthe effect of the segregated network on the electrical properties of MWCNT and graphene nanosheets (GNS) in high-density polyethylene (HDPE) matrix. The segregated network structure was achieved by alcohol-assisted dispersion and hot-pressing process, which caused the MWCNTs and GNS to be distributed along specific paths, which ultimately resulted in a low electrical percolation threshold. The nanoparticles were first dispersed in alcohol by sonication. The HDPE powder was then blended with the nanoparticle dispersion by sonication. The alcohol was removed, and the dried powder was pressed at room temperature using high pressure. They confirmed that the percolation threshold of MWCNT/HDPE (0.15 vol%) composite was much lower than GNS/HDPE composites (1 vol%). Also, the MWCNT/HDPE composite showed higher electrical conductivity than GNS/HDPE composite at the same filler loadings. Gao et al.[70] also used alcohol-assisted dispersion and hot-press technique to form a segregated network in UHMWPE/MWCNT composites.

Zhang et al.[71] showed that electrically conductive segregated networks could be built in poly(L-lactide)/MWCNT (PLLA/MWCNT) nanocomposites without sacrificing their mechanical properties by simply choosing two PLLA polymers with different viscosities and crystallinities. As shown in Figure 4.9, they first dispersed MWCNTs in PLLA with low viscosity and crystallinity (L-PLLA) to obtain the L-PLLA/MWCNT phase. Next, PLLA particles with high viscosity and crystallinity (H-PLLA) were coated with the L-PLLA/MWCNT phase at 140 °C (below the melting temperature of H-PLLA). The resultant coated H-PLLA particles were compressed above the melting temperature of H-PLLA to form the PLLA/MWCNT nanocomposites having a segregated structure as can be visualized from the AFM scans in Figure 4.10. They achieved an ultralow percolation threshold of 0.019 vol% of MWCNTs. Furthermore, they showed high Young's modulus and tensile strength and maintained high elongation at break due to the formation of continuous and dense MWCNT networks formed by the high interfacial interaction between H-PLLA and L-PLLA, as well as the segregated nature of the structures. Shi et al.[72] demonstrated the success of the same technique using PCL and PLLA.

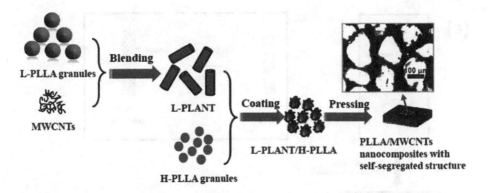

FIGURE 4.9
Schematic Representation of the Procedure for Preparing H-PLLA/L-PLANT Composites with Segregated Structures.

Reprinted from Zhang et al.[71] Copyright 2017 with permission from Royal Society of Chemistry.

George et al.[73] used Vulcastab VL (polyethylene oxide condensate) as a nonionic surfactant to achieve a stable aqueous dispersion of MWCNTs. They blended natural rubber (NR) and MWCNT by latex stage mixing. They found that, instead of being randomly dispersed, MWCNTs were retained at the boundary of rubber particles, resulting in a segregated network as was evident from TEM micrographs. Using this segregated network, they attained a very low electrical percolation threshold of about 0.043 vol%. In a similar study by the same group, the MWCNTs were carboxyl functionalized using H_2SO_4/HNO_3 treatment. The aqueous dispersion of the oxidized MWCNTs was then incorporated in the NR matrix by latex stage mixing as shown in Figure 4.10(a). This again resulted in the oxidized MWCNTs being adhered around NR latex spheres resulting in segregated nanotube network with a very low electrical percolation threshold of 0.086 vol% as can be visualized from Figure 4.10(b).[74] Francis et al.[75] formed a segregated network in polymer latex particles using several nanofillers such as carbon black, indium tin oxide, antimony-doped tin oxide, and CNTs. The latex particles used include poly (vinyl acetate-co-acrylic) polydisperse latex, monodisperse poly (vinyl acetate), and poly (vinyl acetate) polydisperse latex.

4.5 The Organization of MWCNTs during Curing

Well-dispersed suspensions of nanotubes in thermosets have been reported to reaggregate spontaneously because of crosslinking. This is because

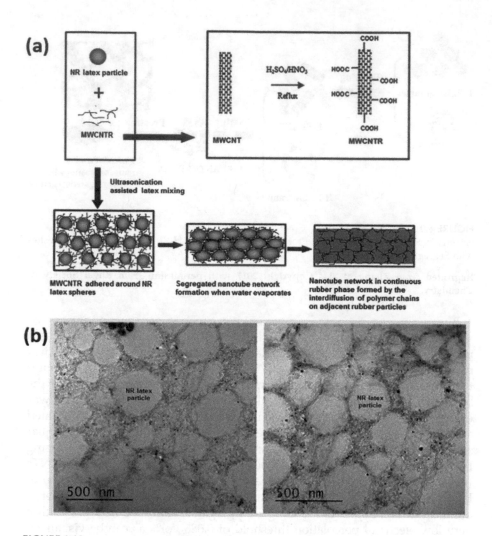

FIGURE 4.10
(a) Schematic for preparation of NR–MWCNT composites with segregated network.(b) TEM images of NR–MWCNTR composite film showing the wrapping of MWCNTs around rubber particles to form a segregated network.

Reprinted from George et al.[74] Copyright 2015 with permission from Elsevier.

nanotubes have a strong tendency to exist in the agglomerated form via their huge surface area, which leads to non-homogeneous dispersion and random distribution of the nanotubes inside the resin. Additionally, a chemical reaction between reactive thermoset precursors and CNTs via suitable functionalization has been found to provide a route for the formation of a well-distributed 3D network for effective stress transfer. It has been reported in the literature that CNTs reduce the glass transition temperature (T_g) of cured thermoset composites due to the inability of the thermoset to achieve complete curing in the presence of CNTs. CNTs cause physical hindrance to the mobility of the active groups in epoxy and the curing agent, thus leading to significantly lower curing temperatures and longer cure times. SEM micrographs by Gerson et al.[76] show that weak bonding of CNTs with the epoxy matrix increases the distance between polymer chains, resulting in a decrease in intermolecular forces, which ultimately translates to lower T_g. The lowering of T_g may be practically important for applications where the thermal properties of the nanocomposites need to be improved.[76–78] The high thermal conductivity of CNTs and the ability of the epoxy resin to disentangle and disperse the CNT bundles offer a higher surface for heat propagation to improve thermal properties.[78,79]

Abdalla et al.[80] studied the effect of fluorinated and carboxylate MWCNTs on the dispersion and cure kinetics of epoxy resin. Comparison of the activation energies, rate constants, gelation behavior, and vitrification times suggested that the cure kinetics of neat resin and fluorinated MWCNT-based sample were similar, but they were different from the sample with carboxylated MWCNTs. The fluorinated MWCNT system was found to be more uniformly dispersed in the matrix; however, heterogeneously dispersed carboxylated MWCNTs hindered the mobility of the reactive moieties in the resin inducing disruption in the reaction stoichiometry on the local scale and altering cure kinetics.

Wardle et al.[81] developed a novel method to obtain an aligned epoxy-MWCNT composite. They first induced mechanical densification of vertically aligned CNT forests. Furthermore, the epoxy was infused in it using capillarity-induced wetting. They demonstrated that aligned epoxy/MWCNT nanocomposite could be fabricated to up to 22% volume fractions, where inter-CNT spacing approaches characteristic lengths of polymer chains.

4.6 Phase Separation of Polymer Blends as a Tool in Organizing CNT Network Structures

The blending of two polymers is one of the most effective and cheapest ways of obtaining materials that have improved mechanical properties.[82] When two polymers are mixed together, they can either be miscible or immiscible depending on factors governing thermodynamics. Low molecular weight polymers are usually miscible due to the combinatorial entropy that runs high in such systems when mixed. The Gibbs free energy of mixing should be less than zero (negative) to favor miscibility, as according to Equation 4.1.

$$\Delta G_m = \Delta H_m - T \Delta S_m \tag{4.1}$$

where ΔG_m represents Gibbs free energy of mixing, ΔH_m is the enthalpy of mixing, and ΔS_m is the entropy of mixing. The second derivative of the Gibbs free energy should be greater than zero to further satisfy miscibility.

$$\left(\frac{\partial^2 \Delta G_m}{\partial \Phi_i^2} \right)_{T,P} > 0 \tag{4.2}$$

For high molecular weight, the combinatorial entropy and increase in temperature will not favor much miscibility, henceforth other factors such as enthalpy of mixing and non-combinatorial entropies will favor miscibility. Phase separation of miscible polymer blends usually occurs via spinodal decomposition or nucleation and growth mechanism.[83,84] In other polymer blends such as PS/PVME phase separation also occurs via viscoelastic phase separation.[85] Spinodal decomposition usually takes place in near critical compositions where there are large-scale concentration fluctuations. This type of phase separation mechanism usually transpires in the unstable regime of the phase separation diagram, mainly due to concentration fluctuations developing into phase-separated regions. Nucleation and growth usually occur in off-critical compositions; this type of phase separation usually results in a droplet-matrix morphology, while phase separation that occurs through spinodal decomposition usually yields a co-continuous morphology. Viscoelastic phase separation occurs in dynamic asymmetrical polymer blends, which are polymer blends having a large difference in their glass transition temperatures.[86,87] This type of phase separation leads to an interaction of networks, usually in polymer blends, the minor phase form droplets leading to droplet matrix morphology but in viscoelastic phase separation, the minor phase results in a network structure.

Spinodal decomposition is the most favorable form of phase separation because it allows the alignment of CNTs in one continuous phase. Bose et al.[88] were able to use phase separation of PαMSAN/PMMA as a tool in dispersing of MWCNT. When two partially miscible polymers with nanoparticles phase-separate though spinodal decomposition, the nanoparticles can either be in one of the phases of the polymers or at the interphase. Localization of nanoparticles in polymer blends is mainly governed by thermodynamics, kinetics, and processing conditions.[89] Secondary agglomeration of CNTs is one of the main challenges that is encountered in organizing CNTs inside a host matrix. This is mainly due to the van der Waals forces of attraction between the nanoparticles.[90,91] To overcome these challenges, CNTs are usually chemically modified on their surfaces to enhance a perfect organization in the host matrix.[92] Owing to the absence of functional groups on their surface, CNTs are chemically treated with oxidizing agents such as nitric acid or potassium permanganate to induce oxygen-containing groups on their surface.[93] There are two ways of covalently attaching polymers on CNTs: "grafting to" and "grafting from" methods.[94–96] In the former method, the functional groups of the polymer chains react with the functional groups present on CNT surface and in the latter method, polymerization of the polymer chains is originated on the surface of the CNTs. The main advantage of the "grafting from" over the "grafting to" method is that high grafting density is achieved. There are multiple ways that have been generated to covalently functionalize the surface of CNTs with polymer chains. Kar et al.[97] successfully managed to graft polystyrene on CNTs by nitrene chemistry. They then mixed the polymer-grafted CNTs into PS/PVME blend solution using a shear mixer for 45 min at 8000 rpm. After annealing the blends with the polymer-grafted nanoparticles for 2 h at 125 °C, they were able to observe a good dispersion of the polymer-grafted CNTs because of the polymer chains, which prevented the agglomeration of CNTs, hence perfectly organizing the nanoparticles throughout the PVME matrix. Phase separation henceforth can be used as an effective tool in organizing perfect networks of nanoparticles such as CNTs.

CNTs at the polymer blend interface can also result in an organized network structure. In immiscible polymer blends, chemical modification and processing factors are some of the main methods that can be used to localize nanoparticles at the interphase. Bose et al.[98] were able to jam CNTs at the interphase of PA6/ABS polymer blend system by chemically modifying MWCNT with SMA. Baudouin etal.[99] managed to localize unfunctionalized CNTs at the interphase of PA/EA (Figure 4.11) by first mixing the nanoparticles with PA6, allowing them to be dispersed, and then

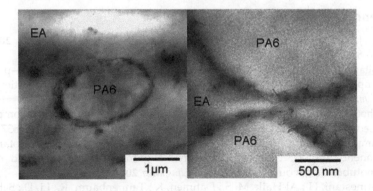

FIGURE 4.11
TEM micrographs showing localization of MWCNT at the interphase of PA/EA polymer blend system.

Reprinted from Baudouin et al.[99] Copyright 2010 with permission from Elsevier.

adding EA through melt-mixing processing. Polymer blend systems play a major role in creating a perfect network structure of CNTs.

4.7 Conclusions

CNTs can be perfectly organized during crystallization of polymers through the formation of shish–kebab morphology. This method is favorable because it does not disrupt the chemical structure of CNTs hence they donot lose their intrinsic properties. Block copolymers can also play an important role in organizing a CNT network. The ability of crystalline block copolymers to decorate themselves along the CNTs enables a periodic patterning of CNTs. To achieve good interphase connectivity at low nanoparticle loading, one of the common techniques that can be used is the role of segregated networks. This technique usually leads to low percolation thresholds in polymers. CNTs can also be perfectly organized into network structures during curing. This is mainly due to the reaction between the thermoset precursor and the CNTs through suitable functionalization resulting in a well-distributed 3D network structure. Phase separation of miscible polymer blends also serves as an effective mechanism in the formation of an organized CNT network structure. Phase separation of polymer blends via spinodal decomposition allows the CNTs to be localized either in one of the polymer phases or at the interphase. This chapter highlights the various strategies that result in "organization of CNTs" within polymeric systems on account of polymer crystallization, crosslinking, phase separation, and will help guide researchers working in this field from both academia and industry.

References

1. Jog, J., Crystallisation in polymer nanocomposites. *Mater. Sci. Technol.* **2006**, 22, 797–806.
2. Spitalsky, Z.; Tasis, D.; Papagelis, K.; Galiotis, C., Carbon nanotube–polymer composites: chemistry, processing, mechanical and electrical properties. *Prog. Polym. Sci.* **2010**, 35, 357–401.
3. Sahoo, N. G.; Rana, S.; Cho, J. W.; Li, L.; Chan, S. H., Polymer nanocomposites based on functionalized carbon nanotubes. *Prog. Polym. Sci.* **2010**, 35, 837–867.
4. Camponeschi, E.; Florkowski, B.; Vance, R.; Garrett, G.; Garmestani, H.; Tannenbaum, R., Uniform directional alignment of single-walled carbon nanotubes in viscous polymer flow. *Langmuir* **2006**, 22, 1858–1862.
5. Garmestani, H.; Al-Haik, M. S.; Dahmen, K.; Tannenbaum, R.; Li, D.; Sablin, S. S.; Hussaini, M. Y., Polymer-mediated alignment of carbon nanotubes under high magnetic fields. *Adv. Mater.* **2003**, 15, 1918–1921.
6. Pradhan, B.; Kohlmeyer, R. R.; Chen, J., Fabrication of in-plane aligned carbon nanotube–polymer composite thin films. *Carbon* **2010**, 48, 217–222.

7. Xie, X.-L.; Mai, Y.-W.; Zhou, X.-P., Dispersion and alignment of carbon nanotubes in polymer matrix: a review. *Mater. Sci. Eng. R Rep.* **2005**, 49, 89–112.
8. Ahir, S. V.; Huang, Y. Y.; Terentjev, E. M., Polymers with aligned carbon nanotubes: active composite materials. *Polymer* **2008**, 49, 3841–3854.
9. Goh, P.; Ismail, A.; Ng, B., Directional alignment of carbon nanotubes in polymer matrices: contemporary approaches and future advances. *Composites, Part A* **2014**, 56, 103–126.
10. Yuan, Q.; Misra, R., Polymer nanocomposites: current understanding and issues. *Mater. Sci. Technol.* **2006**, 22, 742–755.
11. Liu, Y.; Kumar, S., Polymer/carbon nanotube nano composite fibers–a review. *ACS Appl. Mater. Interfaces* **2014**, 6, 6069–6087.
12. Bhattacharyya, A. R.; Sreekumar, T. V.; Liu, T.; Kumar, S.; Ericson, L. M.; Hauge, R. H.; Smalley, R. E., Crystallization and orientation studies in polypropylene/single wall carbon nanotube composite. *Polymer* **2003**, 44, 2373–2377.
13. Valentini, L.; Biagiotti, J.; Kenny, J. M.; Santucci, S., Morphological characterization of single-walled carbon nanotubes-PP composites. *Compos. Sci. Technol.* **2003**, 63, 1149–1153.
14. Grady, B. P.; Pompeo, F.; Shambaugh, R. L.; Resasco, D. E., Nucleation of polypropylene crystallization by single-walled carbon nanotubes. *J. Phys. Chem. B* **2002**, 106, 5852–5858.
15. Manchado, M. L.; Valentini, L.; Biagiotti, J.; Kenny, J., Thermal and mechanical properties of single-walled carbon nanotubes–polypropylene composites prepared by melt processing. *Carbon* **2005**, 43, 1499–1505.
16. Sarno, M.; Gorrasi, G.; Sannino, D.; Sorrentino, A.; Ciambelli, P.; Vittoria, V., Polymorphism and thermal behaviour of syndiotactic poly (propylene)/carbon nanotube composites. *Macromol. Rapid Commun.* **2004**, 25, 1963–1967.
17. Sandler, J.; Broza, G.; Nolte, M.; Schulte, K.; Lam, Y.-M.; Shaffer, M., Crystallization of carbon nanotube and nanofiber polypropylene composites. *J. Macromol. Sci., Part B: Phys.* **2003**, 42, 479–488.
18. Probst, O.; Moore, E. M.; Resasco, D. E.; Grady, B. P., Nucleation of polyvinyl alcohol crystallization by single-walled carbon nanotubes. *Polymer* **2004**, 45, 4437–4443.
19. Wanjale, S. D., Crystallization, phase transformation and thermo-mechanical studies of poly (1-butene) nanocomposites. PhD diss., University of Pune **2007**.
20. Anand, K. A.; Jose, T. S.; Agarwal, U.; Sreekumar, T.; Banwari, B.; Joseph, R., PET-SWNT nanocomposite fibers through melt spinning. *Int. J. Polymer. Mater.* **2010**, 59, 438–449.
21. Sandler, J.; Pegel, S.; Cadek, M.; Gojny, F.; Van Es, M.; Lohmar, J.; Blau, W.; Schulte, K.; Windle, A.; Shaffer, M., A comparative study of melt spun polyamide-12 fibres reinforced with carbon nanotubes and nanofibres. *Polymer* **2004**, 45, 2001–2015.
22. Haggenmueller, R.; Fischer, J. E.; Winey, K. I., Single wall carbon nanotube/polyethylene nanocomposites: nucleating and templating polyethylene crystallites. *Macromolecules* **2006**, 39, 2964–2971.
23. Liu, T.; Phang, I. Y.; Shen, L.; Chow, S. Y.; Zhang, W.-D., Morphology and mechanical properties of multiwalled carbon nanotubes reinforced nylon-6 composites. *Macromolecules* **2004**, 37, 7214–7222.

24. Yudin, V. E.; Svetlichnyi, V. M.; Shumakov, A. N.; Letenko, D. G.; Feldman, A. Y.; Marom, G., The nucleating effect of carbon nanotubes on crystallinity in R-BAPB-type thermoplastic polyimide. *Macromol. Rapid Commun.* **2005**, 26, 885–888.

25. Lai, M.; Li, J.; Yang, J.; Liu, J.; Tong, X.; Cheng, H., The morphology and thermal properties of multi-walled carbon nanotube and poly (hydroxybutyrate-co-hydroxyvalerate) composite. *Polym. Int.* **2004**, 53, 1479–1484.

26. Anand, K. A.; Agarwal, U.; Joseph, R., Carbon nanotubes induced crystalliza-tion of poly (ethylene terephthalate). *Polymer* **2006**, 47, 3976–3980.

27. Zhang, S.; Kumar, S., Shaping polymer particles by carbon nanotubes. *Macromol. Rapid Commun.* **2008**, 29, 557–561.

28. Li, L.; Li, C. Y.; Ni, C., Polymer crystallization-driven, periodic patterning on carbon nanotubes. *J. Am. Chem. Soc.* **2006**, 128, 1692–1699.

29. Li, B.; Li, C. Y., Immobilizing Au nanoparticles with polymer single crystals, patterning and asymmetric functionalization. *J. Am. Chem. Soc.* **2007**, 129, 12–13.

30. Li, B.; Ni, C.; Li, C. Y., Poly (ethylene oxide) single crystals as templates for Au nanoparticle patterning and asymmetrical functionalization. *Macromolecules* **2008**, 41, 149–155.

31. Wang, B.; Li, B.; Xiong, J.; Li, C. Y., Hierarchically ordered polymer nanofibers via electrospinning and controlled polymer crystallization. *Macromolecules* **2008**, 41, 9516–9521.

32. Worsley, K. A.; Moonoosawmy, K. R.; Kruse, P., Long-range periodicity in carbon nanotube sidewall functionalization. *Nano Lett.* **2004**, 4, 1541–1546.

33. De Gennes, P.-G., *Simple Views on Condensed Matter*. World Scientific, Singapore **1998**.

34. Kimata, S.; Sakurai, T.; Nozue, Y.; Kasahara, T.; Yamaguchi, N.; Karino, T.; Shibayama, M.; Kornfield, J. A., Molecular basis of the shish-kebab morphology in polymer crystallization. *Science* **2007**, 316, 1014–1017.

35. Li, C. Y.; Li, L.; Cai, W.; Kodjie, S. L.; Tenneti, K. K., Nanohybrid shish-kebabs: periodically functionalized carbon nanotubes. *Adv. Mater.* **2005**, 17, 1198–1202.

36. Kodjie, S. L.; Li, L.; Li, B.; Cai, W.; Li, C. Y.; Keating, M., Morphology and crystallization behavior of HDPE/CNT nanocomposite. *J. Macromol. Sci., Part B: Phys.* **2006**, 45, 231–245.

37. Geil, P., *Polymer Single Crystals*. Robert Krieger Pub., Huntington, NY **1973**.

38. Pennings, A., Bundle-like nucleation and longitudinal growth of fibrillar poly-mer crystals from flowing solutions. *J. Polym. Sci., Polym. Symp.* **1977**, 59, 55–86.

39. Tuinstra, F.; Baer, E., Epitaxial crystallization of polyethylene on graphite. *Journal of Polymer Science Part B: Polymer Letters.* **1970**, 8, 861–865.

40. Takenaka, Y.; Miyaji, H.; Hoshino, A.; Tracz, A.; Jeszka, J. K.; Kucinska, I., Interface structure of epitaxial polyethylene crystal grown on HOPG and MoS2 substrates. *Macromolecules* **2004**, 37, 9667–9669.

41. Zhang, Q.; Lippits, D. R.; Rastogi, S., Dispersion and rheological aspects of SWNTs in ultrahigh molecular weight polyethylene. *Macromolecules* **2006**, 39, 658–666.

42. Uehara, H.; Kakiage, M.; Sekiya, M.; Sakuma, D.; Yamonobe, T.; Takano, N.; Barraud, A.; Meurville, E.; Ryser, P., Size-selective diffusion in nanoporous but flexible membranes for glucose sensors. *ACS Nano* **2009**, 3, 924–932.

43. Zhang, Z.; Xu, Q.; Chen, Z.; Yue, J., Nanohybrid shish-kebabs: supercritical CO_2-induced PE epitaxy on carbon nanotubes. *Macromolecules* **2008**, 41, 2868–2873.

44. Cai, W.; Li, C. Y.; Li, L.; Lotz, B.; Keating, M.; Marks, D., Submicrometer scroll/tubular lamellar crystals of Nylon 6, 6. *Adv. Mater.* **2004**, 16, 600–605.
45. Li, L.; Li, C. Y.; Ni, C.; Rong, L.; Hsiao, B., Structure and crystallization behavior of Nylon 66/multi-walled carbon nanotube nanocomposites at low carbon nanotube contents. *Polymer* **2007**, 48, 3452–3460.
46. Li, L.; Yang, Y.; Yang, G.; Chen, X.; Hsiao, B. S.; Chu, B.; Spanier, J. E.; Li, C. Y., Patterning polyethylene oligomers on carbon nanotubes using physical vapor deposition. *Nano Lett.* **2006**, 6, 1007–1012.
47. Rao, S. G.; Huang, L.; Setyawan, W.; Hong, S., Large-scale assembly of carbon nanotubes. *Nature* **2003**, 425, 36.
48. Ahn, J.-H.; Kim, H.-S.; Lee, K. J.; Jeon, S.; Kang, S. J.; Sun, Y.; Nuzzo, R. G.; Rogers, J. A., Heterogeneous three-dimensional electronics by use of printed semiconductor nanomaterials. *Science* **2006**, 314, 1754.
49. Kang, S. J.; Kocabas, C.; Ozel, T.; Shim, M.; Pimparkar, N.; Alam, M. A.; Rotkin, S. V.; Rogers, J. A., High-performance electronics using dense, perfectly aligned arrays of single-walled carbon nanotubes. *Nat. Nanotechnol.* **2007**, 2, 230.
50. Li, X.; Zhang, L.; Wang, X.; Shimoyama, I.; Sun, X.; Seo, W.-S.; Dai, H., Langmuir–blodgett assembly of densely aligned single-walled carbon nanotubes from bulk materials. *J. Am. Chem. Soc.* **2007**, 129, 4890–4891.
51. LeMieux, M. C.; Roberts, M.; Barman, S.; Jin, Y. W.; Kim, J. M.; Bao, Z.; Self-Sorted, A., Nanotube networks for thin-film transistors. *Science* **2008**, 321, 101.
52. Kam, N. W. S.; Connell, M.; Wisdom, J. A.; Dai, H., Carbon nanotubes as multifunctional biological transporters and near-infrared agents for selective cancer cell destruction. *Proc. Natl. Acad. Sci. U.S.A.* **2005**, 102, 11600.
53. Czerw, R.; Guo, Z.; Ajayan, P. M.; Sun, Y.-P.; Carroll, D. L., Organization of polymers onto carbon nanotubes: a route to nanoscale assembly. *Nano Lett.* **2001**, 1, 423–427.
54. Zheng, M.; Jagota, A.; Strano, M. S.; Santos, A. P.; Barone, P.; Chou, S. G.; Diner, B. A.; Dresselhaus, M. S.; McLean, R. S.; Onoa, G. B.; Samsonidze, G. G.; Semke, E. D.; Usrey, M.; Walls, D. J., Structure-based carbon nanotube sorting by sequence-dependent DNA assembly. *Science* **2003**, 302, 1545.
55. Richard, C.; Balavoine, F.; Schultz, P.; Ebbesen, T. W.; Mioskowski, C., Supramolecular self-assembly of lipid derivatives on carbon nanotubes. *Science* **2003**, 300, 775.
56. Li, B.; Li, L.; Wang, B.; Li, C. Y., Alternating patterns on single-walled carbon nanotubes. *Nat. Nanotechnol.* **2009**, 4, 358.
57. Li, Z.; Kesselman, E.; Talmon, Y.; Hillmyer, M. A.; Lodge, T. P., Multicompartmentmicelles from ABC miktoarm stars in water. *Science* **2004**, 306, 98–101.
58. Cui, H.; Chen, Z.; Zhong, S.; Wooley, K. L.; Pochan, D. J., Block copolymer assembly via kinetic control. *Science* **2007**, 317, 647–650.
59. Yu, N.; Zheng, X.; Xu, Q.; He, L., Controllable-induced crystallization of PE-b-PEO on carbon nanotubes with assistance of supercritical CO2: effect of solvent. *Macromolecules* **2011**, 44, 3958–3965.
60. Park, H. S.; Choi, B. G.; Hong, W. H.; Jang, S.-Y., Interfacial interactions of single-walled carbon nanotube/conjugated block copolymer hybrids for flexible transparent conductive films. *J. Phys. Chem. C* **2012**, 116, 7962–7967.
61. Wang, W.; Laird, E. D.; Li, B.; Li, L.; Li, C. Y., Tuning periodicity of polymer-decorated carbon nanotubes. *Sci. China: Chem.* **2012**, 55, 802–807.

62. Chen, W.; Tao, X., Self-organizing alignment of carbon nanotubes in thermoplastic polyurethane. *Macromol. Rapid Commun.* **2005**, 26, 1763–1767.

63. Smalley, R.; Colbert, D., *Self Assembly of Fullerene Tubes and Balls, Talk to: Robert A. Welch Foundation, Houston, TX, October* **1995**.

64. Ruoff, R. S.; Lorents, D. C., Mechanical and thermal properties of carbon nanotubes. *Carbon* **1995**, 33, 925–930.

65. Salvetat, J.-P.; Briggs, G. A. D.; Bonard, J.-M.; Bacsa, R. R.; Kulik, A. J.; Stöckli, T.; Burnham, N. A.; Forró, L., Elastic and shear moduli of single-walled carbon nanotube ropes. *Phys. Rev. Lett.* **1999**, 82, 944.

66. Bouchet, J.; Carrot, C.; Guillet, J.; Boiteux, G.; Seytre, G.; Pineri, M., Conductive composites of UHMWPE and ceramics based on the segregated network concept. *Polym. Eng. Sci.* **2000**, 40, 36–45.

67. Mierczynska, A.; Friedrich, J.; Maneck, H.; Boiteux, G.; Jeszka, J., Segregated network polymer/carbon nanotubes composites. *Open Chem.* **2004**, 2, 363–370.

68. Lisunova, M. O.; Mamunya, Y. P.; Lebovka, N. I.; Melezhyk, A. V., Percolation behaviour of ultrahigh molecular weight polyethylene/multi-walled carbon nanotubes composites. *Eur. Polym. J.* **2007**, 43, 949–958.

69. Du, J.; Zhao, L.; Zeng, Y.; Zhang, L.; Li, F.; Liu, P.; Liu, C., Comparison of electrical properties between multi-walled carbon nanotube and graphene nanosheet/high density polyethylene composites with a segregated network structure. *Carbon* **2011**, 49, 1094–1100.

70. Gao, J.-F.; Li, Z.-M.; Meng, Q.-J.; Yang, Q., CNTs/UHMWPE composites with a two-dimensional conductive network. *Mater. Lett.* **2008**, 62, 3530–3532.

71. Zhang, K.; Li, G.-H.; Feng, L.-M.; Wang, N.; Guo, J.; Sun, K.; Yu, K.-X.; Zeng, J.-B.; Li, T.; Guo, Z.; Wang, M., Ultralow percolation threshold and enhanced electromagnetic interference shielding in poly(L-lactide)/multi-walled carbon nanotube nanocomposites with electrically conductive segregated networks. *J. Mater. Chem. C* **2017**, 5, 9359–9369.

72. Shi, Y.-D.; Lei, M.; Chen, Y.-F.; Zhang, K.; Zeng, J.-B.; Wang, M., Ultralow percolation threshold in poly(L-lactide)/poly(ε-caprolactone)/multiwall carbon nanotubes composites with a segregated electrically conductive network. *J. Phys. Chem. C* **2017**, 121, 3087–3098.

73. George, N.; Bipinbal, P.; Bhadran, B.; Mathiazhagan, A.; Joseph, R., Segregated network formation of multiwalled carbon nanotubes in natural rubber through surfactant assisted latex compounding: a novel technique for multifunctional properties. *Polymer* **2017**, 112, 264–277.

74. George, N.; Julie Chandra, C. S.; Mathiazhagan, A.; Joseph, R., High performance natural rubber composites with conductive segregated network of multiwalled carbon nanotubes. *Compos. Sci. Technol.* **2015**, 116, 33–40.

75. Francis, L. F.; Grunlan, J. C.; Sun, J.; Gerberich, W. W., Conductive coatings and composites from latex-based dispersions. *Colloids Surf. Physicochem. Eng. Aspects* **2007**, 311, 48–54.

76. Gerson, A. L.; Bruck, H. A.; Hopkins, A. R.; Segal, K. N., Curing effects of single-wall carbon nanotube reinforcement on mechanical properties of filled epoxy adhesives. *Composites, Part A* **2010**, 41, 729–736.

77. Tao, K.; Yang, S.; Grunlan, J. C.; Kim, Y.-S.; Dang, B.; Deng, Y.; Thomas, R. L.; Wilson, B. L.; Wei, X., Effects of carbon nanotube fillers on the curing processes of epoxy resin-based composites. *J. Appl. Polym. Sci.* **2006**, 102, 5248–5254.

78. Puglia, D.; Valentini, L.; Armentano, I.; Kenny, J. M., Effects of single-walled carbon nanotube incorporation on the cure reaction of epoxy resin and its detection by Raman spectroscopy. *Diamond Relat. Mater.* **2003**, 12, 827–832.
79. Liu, L.; Ye, Z., Effects of modified multi-walled carbon nanotubes on the curing behavior and thermal stability of boron phenolic resin. *Polym. Degrad. Stab.* **2009**, 94, 1972–1978.
80. Abdalla, M.; Dean, D.; Robinson, P.; Nyairo, E., Cure behavior of epoxy/ MWCNT nanocomposites: the effect of nanotube surface modification. *Polymer* **2008**, 49, 3310–3317.
81. Wardle, B. L.; Saito, D. S.; García, E. J.; Hart, A. J.; de Villoria, R. G.; Verploegen, E. A., Fabrication and characterization of ultrahigh-volume-fraction aligned carbon nanotube–polymer composites. *Adv. Mater.* **2008**, 20, 2707–2714.
82. Han, D.; Wen, T. J.; Han, G.; Deng, Y.-Y.; Deng, Y.; Zhang, Q.; Fu, Q., Synthesis of Janus POSS star polymer and exploring its compatibilization behavior for PLLA/PCL polymer blends. *Polymer* **2018**, 136, 84–91.
83. Muzata, T. S.; Jagadeshvaran, P.; Kar, G. P.; Bose, S., Phase miscibility and dynamic heterogeneity in PMMA/SAN blends through solvent free reactive grafting of SAN on graphene oxide. *Phys. Chem. Chem. Phys.* **2018**, 20, 19470–19485.
84. Yeganeh, J. K.;., Dynamics of nucleation and growth mechanism in the presence of nanoparticles or block copolymers: polystyrene/poly (vinyl methyl ether). *Polym. Bull.* **2018**, 75, 1–15.
85. Xavier, P.; Mapping the transient morphologies and demixing behavior of polystyrene/poly (vinyl methyl ether) blend in the presence of multiwall carbon nanotubes. PhD diss., Indian Institute of Science Bangalore, **2018**.
86. Xavier, P.; Nair, K. M.; Lasitha, K.; Bose, S., Is kinetic polymer arrest very specific to multiwalled carbon nanotubes? *Phys. Chem. Chem. Phys. Phys. Chem. Chem. Phys.* **2016**, 18, 29226–29238.
87. Xavier, P.; Bharati, A.; Madras, G.; Bose, S., An unusual demixing behavior in PS–PVME blends in the presence of nanoparticles. *Phys. Chem. Chem. Phys.* **2014**, 16, 21300–21309.
88. Bose, S.; Ozdilek, C.; Leys, J.; Seo, J. W.; Wubbenhorst, M.; Vermant, J.; Moldenaers, P., Phase separation as a tool to control dispersion of multiwall carbon nanotubes in polymeric blends. *ACS Appl. Mater. Interfaces* **2010**, 2, 800–807.
89. Pawar, S. P.; Bose, S., Peculiar morphological transitions induced by nanoparticles in polymeric blends: retarded relaxation or altered interfacial tension? *Phys. Chem. Chem. Phys.* **2015**, 17, 14470–14478.
90. Hassanzadeh-Aghdam, M. K.; Mahmoodi, M. J.; Ansari, R., Creep performance of CNT polymer nanocomposites-An emphasis on viscoelastic interphase and CNT agglomeration. *Composites, Part B* **2019**, 168, 274–281.
91. Yang, Q. S.; He, X. Q.; Liu, X.; Leng, F. F.; Mai, Y.-W., The effective properties and local aggregation effect of CNT/SMP composites. *Composites, Part B* **2012**, 43, 33–38.
92. Kar, G. P.; Bose, S., Nucleation barrier, growth kinetics in ternary polymer blend filled with preferentially distributed carbon nanotubes. *Polymer* **2017**, 128, 229–241.
93. Lima, A. P.; Catto, A. C.; Longo, E.; Nossol, E.; Richter, E. M.; Munoz, R. A., Investigation on acid functionalization of double-walled carbon nanotubes of

different lengths on the development of amperometric sensors. *Electrochim. Acta* **2019**, 299, 762–771.

94. Qin, S.; Qin, D.; Ford, W. T.; Resasco, D. E.; Herrera, J. E., Functionalization of single-walled carbon nanotubes with polystyrene via grafting to and grafting from methods. *Macromolecules* **2004**, 37, 752–757.

95. Gan, L.; Shang, S.; Mei, C.; Xu, L.; Tan, L.; Hu, E., Wet functionalization of carbon nanotubes and its applications in rubber composites. In *Carbon-Based Nanofillers and Their Rubber Nanocomposites*, Editor(s): Yaragalla S., Mishra R., Thomas S., Kalarikkal N., Maria H. J., Elsevier, **2019**.

96. Ma, W.; Zhao, Y.; Zhu, Z.; Guo, L.; Cao, Z.; Xia, Y.; Yang, H.; Gong, F.; Zhong, J., Synthesis of poly (methyl methacrylate) grafted multiwalled carbon nanotubes via a combination of RAFT and alkyne-azide click reaction. *Appl. Sci.* **2019**, 9, 603.

97. Kar, G. P.; Xavier, P.; Bose, S., Polymer-grafted multiwall carbon nanotubes functionalized by nitrene chemistry: effect on cooperativity and phase miscibility. *Phys. Chem. Chem. Phys.* **2014**, 16, 17811–17821.

98. Bose, S.; Bhattacharyya, A. R.; Kodgire, P. V.; Misra, A.; Pötschke, P., Rheology, morphology, and crystallization behavior of melt-mixed blends of polyamide6 and acrylonitrile-butadiene-styrene: influence of reactive compatibilizer premixed with multiwall carbon nanotubes. *J. Appl. Polym. Sci.* **2007**, 106, 3394–3408.

99. Baudouin, A.-C.; Devaux, J.; Bailly, C., Localization of carbon nanotubes at the interface in blends of polyamide and ethylene–acrylate copolymer. *Polymer* **2010**, 51, 1341–1354.

5

Mechanical Properties of CNT Network-Reinforced Polymer Composites

Sushant Sharma, Abhishek Arya, Sanjay R. Dhakate, and Bhanu Pratap Singh

CSIR-National Physical Laboratory
New Delhi, India

Academy of Scientific & Innovative Research (AcSIR)
Ghaziabad, Uttar Pradesh, India

5.1 Introduction

Carbon nanotube (CNT), a fascinating material with outstanding properties, has inspired the researchers because of its wide range of potential applications [1]. The theoretical and experimental studies concluded that CNTs possess low density, high aspect ratio, and unique mechanical properties, which make them particularly attractive for use as reinforcement in composite materials [2–5]. Till date, thousands of publications have reported on the numerous aspects of enhancement in mechanical properties of different types of polymer systems through the reinforcement of CNTs. These studies involve the effect of various parameters such as type, growth mechanism, chemical treatment of CNT used, CNT loading, type of matrix system, processing techniques, and orientation.

Outstanding electrical, thermal, and mechanical properties of CNTs are already explored, but expected outcomes of their reinforcement are not yet fully realized [6–8]. Traditionally, CNTs are reinforced as discontinuous dispersion form in nanocomposites, but their effective dispersion in the polymer matrix is a challenging task. CNTs have a strong tendency to re-agglomerate during the processing of composite materials due to their extremely high aspect ratio and strong van der Waals interaction. Therefore, desired chemical treatment is provided to modify the surface of CNTs and hence to obtain the better nanotube dispersion [9–11]. However, chemical treatment damages the crystalline structure of CNTs and reduces their strength and modulus [12]. Thus, in situ polymerization is an improved method to disperse the CNTs [13]. In one of the study by Jiang et al., polyimide (PI)–CNT composites were fabricated by in situ polymerization, which effectively increased the electrical and mechanical properties of reinforced composite [14]. Further, some investigators utilized the surfactant-assisted dispersion technique to improve the dispersion efficiency [15], while some used the chemical functionalization followed by surfactant dispersion [16], but were unable to obtain the high nanofiller loading. Commercially available CNTs are mostly in bundle form, especially chemical vapor grown CNTs, which are in highly aggregated and entangled state. These aggregated bundles of CNTs are unable to give more than 5% loading by conventionally used composite processing techniques such as solvent casting, resin transfer molding, and injection molding. Beyond certain weight percent, not only the properties deteriorate, but increased viscosity of polymer also reduces the processability of composite fabrication.

It is commonly observed that the mechanical, electrical, and thermal properties of composites strongly depend on the filler content. High loading of CNTs allows the CNT properties to dominate those of their composites. Thus, it is important to fabricate high loading of CNT-reinforced polymer composites. To overcome the problems associated with conventional processing methods, several techniques have been developed to achieve the high loading (10–50%) of CNT-reinforced composites, which include hot-press molding, mechanical densification, layer by layer method, resin infusion technique, and so on [17–23].

In the aforementioned techniques, thin sheets of random or aligned CNT network are prepared by various techniques and followed by resin infiltration. This self-assembled thin network of CNT is called "carbon paper" or "bucky paper." However, alignment of CNT has played an important role in improving mechanical and other functional properties of composites that have been accomplished by various external agents and in situ techniques.

Therefore, this chapter summarizes the fabrication techniques of CNT sheets and their reinforced composites. The effect of various CNT orientation techniques on mechanical properties of nanocomposite is studied in detail, which includes the alignment technique used in low and high CNT loading condition. The mechanical properties of nanocomposites with respect to orientation of CNT are also discussed. Finally, different applications of these CNT network-reinforced composites are discussed, which cover the present and future scenario of CNT-reinforced nanocomposites.

5.2 CNT Sheet

CNT sheet or bucky paper is a macroscopic free-standing porous network of CNT in which CNTs are either randomly or directionally oriented. It has many potential applications including actuator, capacitor, electrodes, field emission devices, and multifunctional structure [24–32]. These thin porous networks are usually made using the ideology of an ancient paper-making technique that requires pulp. This pulp is nothing but an entangled aggregate of a few millimeter long cellulose fibers, which is filtered and dried to fabricate the paper. Similarly, bucky paper also requires long length and high purity CNTs, which are generally obtained by chemical vapor deposition (CVD) technique [33,34]. The suspension of aggregated CNTs is filtered and dried to make a bucky paper. The bucky paper prepared by this technique contains randomly oriented CNTs, which are not helpful in realizing the real potential of CNTs. Therefore, various techniques are used later on to prepare the oriented CNT sheet, which will be discussed in the next section.

5.3 Fabrication of CNT Sheet

5.3.1 Randomly Oriented CNT Sheet

The first step is to prepare the free-standing network of CNT in the form of sheet or paper involves the production of long-length (100 µm to few mm) CNTs having high aspect ratio. These CNTs can be made into a macroscopic thin-networked sheet using suspension filtration technique, which involves the dispersion of CNT in liquid (water, organic solvent, etc.) by using high energy sonication and centrifugal mixing. The suspension is filtered on porous membrane for removing the liquid and self-assembled network of CNT is then removed gently from the porous membrane. The randomly oriented CNTs are entangled by van der Waals forces. The thickness of the network can be controlled by the concentration of CNT in the suspension. With this technique, transparent (>90% transmittance in a 2–5 mm spectral band) and highly conducting free-standing network of CNT of thickness ~50 nm can be prepared [35]. The main advantage of this technique is simplicity, versatility, and film uniformity. Apart from CNTs, this can be used for other nanosize fibers such as carbon nanofiber, cellulose nanofiber, boron nitride nanotube, and graphene oxide [36–39]. The first CNT network was prepared in 1998 by Smalley et al. in an attempt to test the purity of the single-wall CNTs (SWCNTs) [40]. CNTs were functionalized by using Triton X-100 and then sonicated to make the aqueous suspension. The suspension was filtered through a membrane. The prepared CNT network was only used to check the physical properties of SWCNTs, not for bucky paper. But still, Smalley et al. coined a simple wet processing technique to make randomly oriented bucky paper. Since that time, many more innovative approaches to make bucky paper have been studied and reported.

Earlier conventional Buckner filtration was used to filter the homogeneous suspension of CNTs as depicted in Figure 5.1(a). The homogeneous

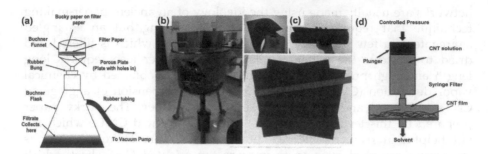

FIGURE 5.1
(a) Buckner filtration assembly, (b) digital image of Buckner filtration assembly for preparing 35 cm × 35 cm randomly oriented bucky paper, (c) 35 cm × 35 cm flexible bucky paper [29] (Reprinted from *Composites Part A: Applied Science and Manufacturing 104*, Sharma S, Singh BP, Chauhan SS, Jyoti J, Arya AK, Dhakate S. Enhanced thermomechanical and electrical properties of multiwalled carbon nanotube paper reinforced epoxy laminar composites, 129–138. Copyright 2018, with permission from Elsevier), (d) schematic of the pressurized filtration process [41] (Reprinted from *Microporous and Mesoporous Materials 184*, Zhang J, Jiang D, Peng H-XA pressurized filtration technique for fabricating carbon nanotube buckypaper: Structure, mechanical and conductive properties, 127–133. Copyright 2014, with permission from Elsevier).

suspension was prepared either by using well-known organic solvents such as acetone, toluene, and dimethyl formamide or by using surfactants such as sodium dodecyl sulfate (SDS), alkali-soluble emulsifier (ASE), Triton X-100 in case of dispersion of CNTs in aqueous medium [42,43]. Our own group is also working in this direction; recently bucky paper of size 35 cm × 35 cm has been prepared by using ASE-refluxed MWCNT by using Buckner filtration assembly (Figure 5.1(b and c)).

The size of the bucky paper is dependent on fabrication assembly but for various advanced applications, there is a stringent requirement of continuous production of bucky paper. Since the technique is simple, the continuous production of sheet is a major challenge in this direction. In Florida State University, continuous production by suspension filtration was performed by using a specially designed filtration device. In that device, continuous deposition of CNTs occurs when long sheet of filtration membrane comes in contact with a rotating filter. This rotating filter moves continuously through a homogeneous CNT suspension and fixes the randomly oriented network of CNTs over the porous membrane [44]. The continuous moving filter technique overcomes the dimensional limitation, which is generally associated with conventional Buckner filtration process for CNTs network. Apart from this, it also limits the changing of filter membrane for the continuous fabrication of CNT paper. Filtration is a very important step in suspension filtration technique because it directly affects the filtration time and efficiency. To reduce the filtration time, various modified techniques of filtration were used. Zhang et al. [41] used pressurized filtration technique to prepare the

CNT bucky paper. In this technique, suspension of CNT was pressurized through a porous membrane to make the uniformly thin bucky paper (Figure 5.1(d)). Similarly, Nanolab, Inc. used a novel hydro-entanglement technique to prepare the bucky paper. In this technique, pressure is applied on filtration membrane by using pressurized water flowing through porous orifice plate. Water jet presses the CNTs over the filter membrane to make a uniform network of CNT [45]. Although these processes are fast and continuous, but they cannot meet the requirement of electronic application, where complex conductive circuits are required. Therefore, an advanced technique is used, in which conductive pattern of CNT is formed over hard or flexible substrate called "Ink jet printing" [46]. In this technique, an ink composed of uniformly dispersed network of CNTs is used to print on a substrate similar to how a printer works. A loaded cartridge of CNT ink is used to print on a running substrate like flexible paper, wafer, or hard substrate. This technique is very effective and can be used to make any complex design in a controlled manner. The speed of fabrication makes ink jet printing, cost effective, and highly scalable. Depending on anticipated application, CNT sheet can also be prepared by using electro-less deposition method. These processing methods produce CNT sheets whose properties are far below the properties of individual CNT due to random fashion of packing. Random orientation causes ineffective contact among the CNT network and causes poor load transfer between them. To overcome this problem by enhancing the tube–tube interaction, several techniques have been used, including surfactant removal, organic solvent densification, cross-linking, oxidation, prior chemical treatment, CNT alignment, and so on. All these mentioned techniques will be discussed in detail in upcoming sections.

5.3.2 Aligned CNT Sheet

It is a big challenge to make a macroscale CNT structure and to fully utilize the inherited properties of CNTs. The first CNT structure was prepared by Zhang et al. [47] and showed the excellent electrical and thermal properties but poor mechanical properties. CNT sheets prepared by the conventional suspension filtration technique commonly have the low strength, low electrical and thermal properties. It is observed that during dispersion, the native structure of individual CNT is disturbed by truncation and inducing many structure defects, which are detrimental to their mechanical properties. Alternatively, the alignment of CNT approach can avoid such problem. In past decades, it is observed that various alignment techniques are developed to fabricate the aligned network of CNT, such as forest spinning, direct spinning, wet spinning, and domino pushing.

In forest spinning technique, a vertically aligned CNT array is directly grown over the pre-deposited catalyst film of high purity by CVD technique (Figure 5.2(a)). This technique requires optimization of various parameters,

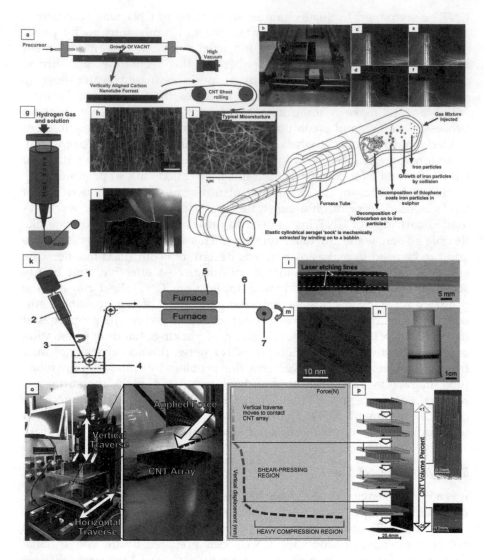

FIGURE 5.2
(a) Schematic showing vertically aligned carbon nanotube (VACNT) growth system, (b) drawing of a multilayer sheet from the side of a CNT Array [48] (Reprinted from *Carbon* 50, no. 11, Pöhls J-H, Johnson MB, White MA, Malik R, Ruff B, Jayasinghe C. Physical properties of carbon nanotube sheets drawn from nanotube arrays, 4175–4183. Copyright 2012, with permission from Elsevier). (c–f) Sequence of consecutive SEM images, given by panels from (c) to (f), of the side view of the CNT forest during the pulling-out process. Red circles show the sequences of points where a pulled-out bundle started pulling the next bundle. Blue circles show the rupture of two adjacent bundles [49] (Reprinted with permission from Kuznetsov AA, Fonseca AF, Baughman RH, Zakhidov AA. Structural model for dry-drawing of sheets and yarns from carbon nanotube forests. *ACS Nano* 5, no. 2:985–993. Copyright 2011 American Chemical Society). (g) Schematic illustration of the synthesis of CNTs and by using

FIGURE 5.2 (CAPTION CONTINUED)

a CVD technique for making continuous CNT fibers. (h) SEM image of as-prepared CNT assembly. (i) Photograph of winding process using a roller (the width of roller is 30 mm) [50] (Reprinted from *Chemical Engineering Journal* 228, Jung Y, Song J, Huh W, Cho D, Jeong Y. Controlling the crystalline quality of carbon nanotubes with processing parameters from chemical vapor deposition synthesis, 1050–1056. Copyright 2013, with permission from Elsevier). (j) Schematic of direct spinning of CNT process with inset shows the typical microstructure of as-prepared CNT [51] (Reprinted from *Extreme Mechanics Letters* 21, Stallard J, Tan W, Smail FR, Gspann T, Boies AM, Fleck NA. The mechanical and electrical properties of direct-spun carbon nanotube mats, 65–75. Copyright 2018, with permission from Elsevier). (k) Schematic of spinning of CNT yarns by the wet spinning method. (l) A long strip of SWCNT array after laser etching. The effective SWCNT array for producing yarns is restricted between the two etching lines and is about 5 mm in width in this figure, (m) transmission electron microscopy (TEM) image of CNT in aligned yarn, (n) 40 m long CNT yarn collected on a winder. The diameter of the yarn is about 10 μm [52], (o) automatic shear press with a graphic illustrating the process of shear pressing CNT arrays, (p) SEM images to the right show the cross-section of an array before (top) and after (bottom) shear pressing [53] (Reprinted from *Composites Part A: Applied Science and Manufacturing* 80, Stahl JJ, Bogdanovich AE, Bradford PD. Carbon nanotube shear-pressed sheet interleaves for Mode I interlaminar fracture toughness enhancement, 127–137. Copyright 2016, with permission from Elsevier).

that is, hydrocarbon type and their flow rate, inert carrier gas, constant growth temperature, uniform thickness of buffer metal oxide layer (Al_2O_3), and uniformly thick catalyst layer (transition element such as Fe, Co, and Ni). Since the thickness of the catalyst is of few nanometers, it remains adhered to substrate and promotes base growth of CNT with narrow diameter distribution, high nucleation density, large aspect ratio, fewer impurities, and better alignment [54]. Aligned and dense bucky paper can be easily prepared by spinning the CNTs from one end and rolling it over a rotating mandrel (Figure 5.2(b)). During spinning, a bundle from edge of the grown substrate is pulled out, which is due to van der Waals forces attracted the adjacent bundle and therefore horizontally aligned CNT yarn is prepared (Figure 5.2(c–f)). The CNTs are tightly packed in alignment direction and provide excellent mechanical, electrical, and thermal properties in the aligned direction. In 2002, Jiang et al. [55] from Tsinghua University proved this concept first time by growing an aligned array of CNT over silicon substrate and drawing it to make yarn of 30 cm long and demonstrated application of CNT filament in bulb. Afterward, lots of research groups start focusing on spinnable aligned forest of CNT and their free standing oriented network for various applications [56–60].

Vertically alignment with high aspect ratio and uniform tube length made it easy for spinning into macroscopic fiber, which after layer by layer deposition form free standing aligned CNT sheets [61]. One of the main advantages of forest spinning is two-step synthesis techniques, in which geometrical parameters of the aligned network can be controlled by controlling the synthesis of CNT. Properties such as length, diameter, straightness, and distance between CNTs can be controlled individually and accurately. The serious downside of this technique is poor productivity, as the size of the aligned network (width and thickness) depends on the size of the CNT array. An effort has been made to overcome this problem by introducing a moving flexible substrate that moves like a conveyor belt [62].

Moving substrate introduces the more complex parameters such as growth time, rotating speed of catalytic substrate, adjustable growth temperature (according to the substrate), and maintenance of low pressure with moving substrate, which causes more complexity in the synthesis. Therefore, these techniques were not capable of fulfilling the industrial scale production and hence single-step technique was introduced, called "direct spinning."

Spinning from a CNT aerogel is a one-step technique, which is capable of fabricating continuous CNT yarn without length limitation (Figure 5.2(g)). The first CNT yarns by aerogel technique were produced jointly by Tsinghua University and Rensselaer Polytechnic Institute in 2002 [63]. The 20 cm long strands of SWCNT were prepared by "enhanced vertical floating technique" using n-hexane as carbon source. The produced strands possess good mechanical (1.2 GPa) and electrical ($1.4–2 \times 10^5$ S/m) properties. In this technique, Ferrocene with thiophene as a catalyst source and ethanol as a carbon source are carried into hot reaction zone by a carrier gas (mixture of argon and hydrogen) [64] through proper control over reaction conditions like catalyst source ratio, carrier gas mixture ratio, and reaction temperature. CNT aerogel can be produced in the hot reaction zone, which twisted and spun at the outlet of the furnace to form the aligned entangled network (Figure 5.2(j)). This technique gained great attention when Cambridge group in 2007 reported the CNT yarn stronger than commercially used carbon fiber [65]. Theoretically, this process can produce the aligned CNT yarn of infinite length as the process is very fast. In this direction, Jung et al. prepared the continuous aligned CNT yarn of high crystallinity and collected it on a 30 mm wide roller (Figure 5.2(h, i)). The improved operating parameters of this technique helped in increasing the length of yarn from 20 cm [63] to more than a kilometer [66] in a few years.

It is found that CNT produced from this method is of large diameter having smaller number of graphitic walls than the CNT produced by forest spinning. These large diameter CNTs do sustain their tubular structural integrity and collapse to form "dog bone" shape at tip [67]. When CNT collapses, the contact area of CNT increases, which enhances the interfacial properties of yarn, therefore improves the mechanical strength of free standing aligned CNT network.

The aforementioned techniques employ solid-state process where CNT can be spun from an aligned array of known dimension or directly from the furnace. The produced aligned CNTs' networks have low packing, sometimes poor orientation and include impurities within their graphitic structure. An alternative fiber production route is wet spinning (Figure 5.2(k)).

Wet spinning can easily scale up to industrial level and is indeed the route to fabricate high-performance fiber such as Kevlar, Twaron, and Thornal carbon fiber [68]. CNT fibers were first time fabricated by this technique in 2000 by Vigolo et al. (Centre de Recherche Paul Pascal, France) [69]. SWCNT produced by electric arc technique was dispersed in aqueous solutions of SDS to form the stable CNT suspension. The SDS surfactant provides the electrostatic repulsion and helped in stabilizing the CNT against van der Waals forces between them. They are thus potentially useful for orienting CNT in the laminar flow. The nanotube suspension was then injected in the co-flowing stream of the polymer solution containing 5 wt% polyvinyl alcohol (PVA) to form the continuous ribbons of aligned CNT network. This standard spinning method was further modified by adopting acid as solvent and coagulants like water [70], for avoiding the surfactants in CNT dispersion and polymer solutions as coagulation bath. The high-quality premade CNTs were dispersed in chlorosulfonic acid at a concentration of 2–6 wt% and filtered to remove the impurities, in order to form spinnable liquid crystal dope, which was extruded through a spinneret (65–130 µm diameter) into a coagulant (acetone or water) bath to remove acid [71,72]. The filaments coming from coagulant bath were collected onto a rotating drum to form a free-standing network of aligned CNT. Liu et al. from Tsinghua University prepared the densely packed CNT yarn by modified wet spinning whose tensile strength was reached up to about 1 GPa (Figure 5.2(k–n)) [52]. For improving the alignment of CNT in fiber, dielectrophoresis and teslaphoresis techniques were implemented, in which CNT was aligned in the presence of electric and magnetic fields, respectively [73,74]. Furthermore, various important modifications have been made in the spinning design to improve the mechanical properties of aligned CNT fiber and free-standing sheets [75].

Aligned CNT paper can also be prepared by rolling the vertically aligned CNT array to change the horizontally aligned sheet. Strong van der Waals interaction between CNT holds the network together, and it can be easily peeled off from beneath substrate [76]. Various techniques using different tools or forces such as rolling [77], surface tension pulling [78], shear pressing [23] were also employed to change the orientation of vertically aligned CNT array. By this similar technique, Stahl et al. prepared horizontally aligned CNT sheet by shear pressing the vertically aligned CNT array and interleaved it for improving the interlaminar shear strength (Figure 5.2(o, p)) of carbon/epoxy laminar composite [53].

Despite outstanding mechanical, electrical, and thermal properties of individual CNT, these properties have not been realized at macroscopic scale. Presently, there are four basic challenges in the field of high-performance CNT network and their reinforced composites: (1) synthesis of high aspect ratio CNT; (2) attaining high volume fraction of CNT while maintaining uniform dispersed network; (3) achieving high degree of CNT alignment along with straightness (4) emerging as industrial viable technique having excellent throughput. These all factors will come into account when researchers start developing free standing oriented CNT network for various applications.

The buckypaper of initially randomly oriented CNTs yields at small stresses of several tens of MPa. But, after achieving the unidirectional orientation, the load carrying capacity in alignment direction is ten times that of randomly distributed CNT bucky paper, which is represented in detail in Tables 5.1 and 5.2. Rashid et al. used high-power tip sonication to prepare the randomly distributed MWCNT bucky papers and tested their mechanical strength. He found that mechanical strength of pristine and functionalized MWCNT bucky paper varied between 1.6 ± 0.7 and 13 ± 2 MPa, respectively [79]. On the other hand, Zhang et al. adopted the two-step technique to prepare the high density aligned CNT bucky paper. In the first step, aligned CNT sheets were prepared by spinning the vertically aligned MWCNT array followed by pressing with load for densification. From the mechanical test, it was found that after pressing, the density of the aligned CNT sheet reached to 1.29 from 0.43 g/cm^3 and with this increased packing, Young's modulus reached to 2190 MPa. The high density aligned CNT sheet not only possesses excellent mechanical property but also has superior electrical and thermal properties. Further, this was employed for light weight heat sink application [80]. Apart from these mentioned studies, there are many other studies available in open literatures that have already proved the improved mechanical and other transport properties in aligned direction compared to randomly oriented bucky paper.

5.4 Functionalization of CNT Sheet

CNTs exhibit high strength and modulus due to a very stable structure consisting of sp^2 bonded carbon atoms. This carbon–carbon double-bonded network of CNT gives stability to CNT, but also limits the realization of many potential applications, where nanotubes need to associate themselves with other polymeric materials. This brings the need of functionalization of CNT.

The challenge in developing the free-standing CNT network-reinforced composite is to create the macroscopic nanocomposite that retains the inherited properties of CNT. In order to achieve the desired properties, establishment of strong chemical affinity with the surrounding polymer is essential. The functionalized CNT sheet interface chemically with polymer and helped in transferring the mechanical and electrical properties effectively. Jiang et al. studied the effect of functionalization on MWCNT-reinforced composite by conducting tensile test on as-prepared and plasma-functionalized MWCNT bucky paper-reinforced composites. He observed that after He/O$_2$ functionalization, tensile strength and Young's modulus improve to

TABLE 5.1

Mechanical properties of randomly oriented CNT bucky paper

S. No.	Type of CNT	Surfactant Used	Technique Used	Tensile Strength (MPa)	Young's Modulus (GPa)	Remarks
1.	SWCNT	—	Suspension filtration	6.49	2.29	SWCNT bucky paper infused with polycarbonate to enhance the electrical and mechanical properties [81]
2.	MWCNT	Triton X-100	Suspension filtration	11.58	1.37	MWCNT bucky paper reinforced polyurethane composite by resin infusion technique [82]
3.	MWCNT	Triton X-100	Suspension filtration	0.82	—	Bucky paper-interleaved composite of polyethylene was treated with microwave irradiation to improve the strength [83]
4.	MWCNT	ASE	Suspension filtration	3.9	0.44	MWCNT was air-oxidized followed by surfactant refluxing [43]
5.	SWCNT	—	Suspension filtration	3.4	0.37	SWCNTs were functionalized to achieve the 3D crosslinking [84]
6.	SWCNT	—	Suspension filtration	10	0.80	SWCNTs were acid-oxidized to improve the dispersion and hence the mechanical properties [47]
7.	SWCNT	—	Suspension filtration	3.6	0.30	SWCNT were dispersed by high-pressure jet mill homogenizer and filtered over cellulose paper [85]
8.	SWCNT	—	Suspension filtration	37	0.95	Chemical modification of SWCNT bucky paper was performed to improve the mechanical properties of pristine bucky [86]
9.	SWCNT	Triton X-100	Suspension filtration	14.2	0.90	SWCNT bucky paper was prepared and external magnetic field (17.3 T) applied to achieve alignment [87]
10.	MWCNT	—	Suspension filtration	14.7	—	MWCNTs were plasma-purified and insitu cross-linked by using 1,4-benzoquinone in methanol [88].

(Continued)

TABLE 5.1 (Cont.)

S. No.	Type of CNT	Surfactant Used	Technique Used	Tensile Strength (MPa)	Young's Modulus (GPa)	Remarks
11.	MWCNT	—	Hydro-entangling suspension filtration	51	10	Pressurized water is used to suppress the CNT membrane and therefore increases the entanglement density [45]
12.	MWCNS	Ethanol/distilled water	Suspension filtration	11.02	2.22	Effect of compression of bucky paper on mechanical, electrical, and thermal properties was studied [89]
13.	Oxidized MWCNT	—	Suspension filtration	4	2	Initially, bucky paper was prepared then soaked in epoxy/harder mixture to fabricate the high loading composite [90]
14.	SWCNT	—	Suspension filtration followed by electrochemical polymerization	68.7	3.25	Individual SWCNT coated with polypyrrole (PPy) by using pulsed electrochemical technique to improve the mechanical and electrical properties [91]
15.	MWCNT	PVP K-30 and Triton X-100	Suspension filtration	1.86	0.36	PVP content was increased to improve the mechanical properties of bucky paper [92]
16.	MWCNT	—	Suspension filtration	3.09	3.75	Mechanical properties of the bucky paper depend on nanoparticle size, lower the size higher the strength [93]
17.	SWCNT	—	Suspension filtration	18.80	6.48	Mechanical properties of the bucky paper depend on nanoparticle size, lower the size higher the strength [93]
18.	SWCNT	Triton X-100	Suspension filtration	11.20	2.15	Ionic liquid soaking by capillary action improves the mechanical properties of bucky paper by liquid bridging [94]

TABLE 5.2

Mechanical properties of oriented CNT bucky paper

S. No.	Method	Carrier Used	Temperature (C)	Densification Applied	Tensile Strength (MPa)	Young's Modulus (GPa)	Remark
1	Spinning of CNT yarns from VACNT	Ar/H$_2$/water vapor/C$_2$H$_2$	750	—	7.9	0.100	Two-step techniques were used to prepare the oriented CNT sheet; first VACNT array was prepared followed by spinning [48]
2	Spinning of CNT yarns from VACNT	Ar/H$_2$/C$_2$H$_2$	750	—	3200	172	In this novel study, micro-combing approach was used before winding the CNT yarn on mandrel [95]
3	Spinning of MWCNT yarns from VACNT	—	—	—	400	—	Plasma functionalized layer by layer deposited MWCNT sheet [96]
4	Direct CVD	—	—	—	>1000	>30	Gas phase pyrolysis was used to synthesize the CNT yarn and sheet with the speed of 150 m/h [97]
5	CVD	—	—	—	44.2	11.9	Noncovalent functionalization was used to improve the crosslinking of CNT [98]
6	FCCVD	Ar/H$_2$	—	Ethanol–water solution	127	3.0	Large-scale production of CNT film [99]
7	FCCVD	Ar/H$_2$	1300	ethanol	243	2.5	Mechanically condensed followed by acid treatment [100]

(Continued)

TABLE 5.2 (Cont.)

S. No.	Method	Carrier Used	Temperature (C)	Densification Applied	Tensile Strength (MPa)	Young's Modulus (GPa)	Remark
8	FCCVD	Ar/H$_2$	—	Ethanol–water solution	90.3	1.2	Pristine CNT sheet was doped with chlorine [101]
9	FCCVD	N$_2$	1200	Ethanol	9600	130	CNT sheet which was further densified [102]
10	FCCVD	Ar/H$_2$	1150–1300	Ethanol	765	—	CNT sheet was densified first by wetting by ethanol, second by rolling and followed by plate pressing [103]
11	FCCVD	Ar/H$_2$	1300	Ethanol–water solution	210	2.6	MWCNT sheet is densified by solution [104]
12	Purchased randomly oriented CNT sheet	—	1200–1300	—	489	13 Gpa	Iodine-doped long-length CNTs were stretched with stretching ratio of 35% for aligning the CNT sheet [105]

30% and 125%, respectively, due to enhanced adhesion between bucky paper and polymer matrix [106]. As Jiang et al. used randomly oriented MWCNTs prepared by vacuum filtration; Malik et al. prepared aligned CNT bucky paper by spinning the aligned MWCNT array and functionalized the MWCNT by plasma treatment before winding over rotating mandrel [96]. These findings and others are perfect example to establish how the functionalized MWCNT network changes their behavior from hydrophobic to hydrophilic and becomes compatible to polymer matrix adhesion.

Generally, two basic techniques are implemented for functionalization of CNTs, which are covalent functionalization and noncovalent functionalization. In covalent functionalization, sp^2-hybridized π orbitals of CNTs provide the possibility for interactions with long conjugate polymer chain [107], while in noncovalent functionalization, physical absorption and wrapping of long polymer chain over CNT surfaces are involved.

5.4.1 Functionalization of Randomly Oriented Sheet

Randomly oriented nature of nanotubes as well as weak van der Waals forces present in the macroscale bucky paper made the mechanical properties unsuitable for advanced applications. To improve the binding nature between individual CNTs, various surface modification techniques are adopted, which include chemical functionalization, air oxidation, e-beam irradiation, plasma treatment, and so on. In chemical functionalization, various chemical groups such as –OH, –COOH, –Cl, and $-NH_3$ are attached covalently over the graphitic wall of CNTs before filtration over permeable membrane [43,108–110]. In e-beam irradiation, chemical cross-linking of CNTs introduced by means of electron beam after filtration and resulting bucky paper showed a significant improvement of mechanical properties without detrimental effect on electrical and thermal properties [111]. Beside e-beam-induced crosslinking, another novel approach is plasma treatment in which functional groups are attached by means of plasma [112]. Plasma treatment overcomes the problem of complicated washing and drying of CNTs for preparing bucky paper. This technique is capable for functionalizing the large amount of CNTs in shorter period of time which is the main reason for growing research in plasma. Plasma treatment gives the freedom to user to control the concentration of functional group over the surface and prevent the saturation of functional groups, which may disturb the electrical and thermal transport properties of CNT. One of the comparative studies between acid-treated and plasma-treated MWCNT is performed by Naseh et al. [113] using temperature-programed desorption technique. He found that acid-treated MWCNTs have more functional groups compared to plasma treated, resulting in less damage to CNT graphitic structure. Additionally, the process is shorter and cleaner than acid functionalization.

5.4.2 Functionalization of Aligned Oriented Sheet

As in the case of randomly oriented CNT bucky paper, chemical functionalization was done prior to filtration on CNTs. In the case of aligned CNT sheet, chemical functionalization can only be performed after the preparation of CNT sheet. But other functionalization techniques such as e-beam irradiations and plasma irradiation can be used during/after the preparation of CNT sheet. In one of the study by Malik et al., CNT sheet was functionalized by using 100 W power He/O$_2$ plasma as shown in Figure 5.3(a, b) [96]. CNT sheets were functionalized by placing plasma torch between rotating cylinder and CNT array, the functionalization rate was 1 cm/s. The CNT sheet was prepared by winding multiple layers (100) of ribbon on the rotating mandrel. By this technique, dense oriented structures of functionalized CNTs were prepared whose tensile strength reached to 400 MPa.

5.5 Reinforcement of CNT Sheet

Since the discovery of CNT, it has been envisaged as great reinforcement for light weight and high-strength composites. The CNTs have been used as reinforcement in various forms, for example, short CNTs with random orientation, aligned CNTs, vertically aligned CNT drawn sheet, and dry spun CNT sheet. Alignment and their internal structure between CNTs play a crucial role for the mechanical strength. Various researches have been reported on this aspect. Theoretically, CNTs have tensile strength of 200 GPa and modulus of 1 TPa [2,114] along with relatively low density, which suits perfectly for light-weight high-strength structural composites. If these

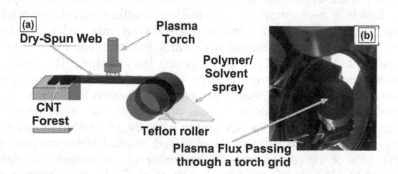

FIGURE 5.3
(a) Schematic of the manufacturing plasma functionalized CNT sheets, (b) single layer of CNT ribbon passing under He/O$_2$ before getting wound on the cylinder [96].

From Malik R, McConnell C, Alvarez NT, Haase M, Gbordzoe S, Shanov V, *RSC Advances* 110, no. 6 (2016): 108840–108850. (Reproduced with permission from The Royal Society of Chemistry.)

inherited properties will be incorporated into polymer matrix composites, the performance of the composites can be reached to the next level.

Dispersion of CNT in polymer matrices is the biggest challenge associated with it; this has been extensively studied for preparing the CNT/polymer nanocomposites. By the conventional techniques, CNT nanocomposites are usually fabricated by dispersing the CNT in suitable polymer by providing suitable solvent or external high-energy agitation. This method involves the efficient dispersion of CNT, thus usually produces the composite with low volume fraction and can only exploit short-length CNTs. This is because long-length CNT possesses a high aspect ratio and tends to agglomerate and tangle during dispersion. Low volume, random-oriented network-reinforced composites are very useful in some electrical and thermal applications such as light weight heat exchangers, electrostatic discharge, and electromagnetic interference-shielding applications [115–117]. For example, Babal et al. utilize the industrial viable technique to disperse the CNT in polycarbonate by using mini extruder with back flow channel and successfully disperse upto 10 wt% of CNT in polycarbonate. After dispersion, mechanical and electrical properties have been studied for electromagnetic interference-shielding applications [118]. Further, for improving dispersion state of CNT, surfactant or functionalizing agent is aided during suspension preparation. Many of the covalent and noncovalent techniques are used to improve the adhesion between CNT and polymer in composite. The short-length CNT-reinforced composites are usually associated with random orientation, which limits the potential. Magnetic and electrostatic forces have been utilized to improve the alignment of the network [119–121]. From the available literature, nanocomposites prepared by solvent-casting and melt-processing techniques show relatively low mechanical properties of randomly oriented CNT polymer composites.

5.5.1 Randomly Oriented CNT Sheet-Reinforced Composite

Bucky papers have gained much attention for the development of CNT composites because their reinforcement provides high-volume fraction of CNTs. Nanocomposites made up of bucky paper and polycarbonate contain ~40–60 wt% of CNTs and offer promising mechanical and electrical properties. The tensile strength and modulus after polycarbonate impregnation reached to 19.65 MPa and 4.22 GPa, while high-loading CNT exhibited electrical conductivity >200 S/cm [81]. Our own group worked on randomly oriented CNT bucky paper-reinforced composites in which high loading (~39 wt%) laminar composites of desired number of plies and thickness have been prepared by vacuum infiltration followed by compression molding. In the first study, epoxy polymer matrix was used to reinforce the bucky paper plies and later on conducting polymer (doped polyaniline) was used [29,124]. Further, these bucky papers were used in carbon fiber-reinforced composites, in which the effect of interlaminar bucky paper reinforcement on lightning strike protection was investigated; it was found that bucky paper applied over the strike face successfully spread the current over the striking face and inhibits the impact of heavy light on underlying carbon fiber layer [125].

5.5.2 Oriented CNT Sheet-Reinforced Composite

Aligned CNT composites possess excellent mechanical and physical properties, can find a wide range of applications such as aerospace, automobile, electronics, and sport goods. Cheng et al. revealed a unique mechanical stretching technique in which they prepared the high wt% (~60 wt%) MWCNT composites with tensile strength and modulus of 2 and 169 GPa, respectively [126]. After functionalization and infiltration with aerospace grade bismaleimide resin, they further successfully improved the strength and modulus to 3 and 350 GPa, respectively [121]. This was the first report which demonstrated that mechanical properties of CNT composite can be more than unidirectional carbon fiber-reinforced composites.

Macroscopically there are two forms of aligned CNT network in which researchers are trying to incorporate the outstanding physical properties of individual CNT: (1) CNT fiber and (2) free-standing CNT sheet. CNT fiber, twisted form of CNT yarn possesses axially aligned and densely packed network of CNT, which exhibit remarkable specific tensile strength and elastic modulus. Recent progress has shown that optimized CNT fiber possesses superior strength of 8.8 GPa, Young's modulus of 357 GPa, and fracture toughness of 121 J/g, which is considerably greater than high-performance industrial fibers [65]. This encouraging performance motivates to reinforce these fibers for structural applications. However, it should be noted that both the strength and modulus of CNT fiber by different groups lie in the wide range of 0.2–8.8 and 30–357 GPa, respectively, irrespective of whether they are synthesized by solid drawing [55,64,126,127] or wet drawing [70,75,128]. It is to be noted that much efforts have been taken on producing the fibers from super aligned CNT array due to high purity, crystallinity, and high-quality long-length CNT. Zhang et al. added one more step of twisting to the drawing process to prepare the CNT fiber of strength and modulus of 330 and 3.3 GPa, respectively [129]. Further, Tran et al. used a modified method of dry spinning, which added several rollers for improving the alignment and polyurethane for densification of CNT fiber, to obtain the strength of 2 GPa [130]. Such fibers are extremely promising elements for reinforcement in CNT fiber composites featuring high strength and electrical conductivity.

CNT fibers have a similar diameter range of well-adopted industrial fibers, and producing 2D reinforcing material out of these 1D CNT fibers is a natural one-step extension. Bogdanovich and group conducted pioneering research on making textile assemblies of CNTs by making 3D braided form of 25 plies of CNT fibers stacking [131,132]. After infusion with epoxy resin, the morphological and mechanical properties were studied and it was observed that epoxy resin penetrate around CNT fiber and also within the fibers. The final composites showed tensile strength of 325 MPa and modulus 24 GPa. CNT sheets drawn from super aligned CNT array can be directly used as reinforcement to prepare high-loading CNT composites. In this regard, Cheng et al. [133] prepared the multi-layer hand drawn free standing sheet and infiltrated it with epoxy resin to prepare the aligned composite of 16.5 wt% having tensile strength of 231.5 MPa and modulus 20.4 GPa. Either in the form of fibers or free standing sheets, densified network is always helpful in improving the mechanical and electrical properties of aligned network. Recently, Xu et al. produced the CNT sheet by solid-state drawing and densified it with a solution to prepare the highly aligned dense network of CNT sheet to prepare the aligned CNT-reinforced composite (Figure 5.4(a–d)) [102]. The study suggests that, apart from alignment, densification also played an important role in improving the mechanical properties of nanocomposite. Figure 5.4(e–j) represents the effect of packing density on mechanical properties of nanocomposites, strength range of 8.0–10.8 GPa with an average of 9.6 GPa, a Young's modulus range of 110–190 GPa with an average of 130 GPa. Similarly, Liu et al. developed the spray-winding technique in which CNT/PVA composites prepared had tensile strength 1.8 GPa, modulus 40–96 GPa, toughness 38–100 J/g, and electrical conductivity 780 S/cm. These composites were stronger than those earlier reported CNT/PVA composites (100–600 MPa) [134–136] and composite fibers (1.5 GPa) [137]. Further, Nam et al. [122] prepared the aligned CNT sheet, first by spinning the aligned CNT array followed by roller densification technique. The mechanically packed aligned CNT sheet was reinforced by using resin transfer compression molding (Figure 5.4(k–p)). The pressed aligned CNT/epoxy composites achieved the highest tensile strength of 526.2 MPa and elastic modulus of 100.2 GPa (Figure 5.4(q, r)). By changing the alignment technique, Garcia et al. [123] prepared horizontally aligned CNT prepreg by mechanical rolling and used it for improving the interlaminar strength (Figure 5.4(s–x)). Compared to regular infiltration technique, which cannot avoid the disruption in the alignment due to flow dynamics of resin, the spray winding technique deposits polymer on each layer without disturbing the alignment of CNT. Secondly, it allows the matrix to penetrate between individual CNT, which ensures the effective load transfer between the CNTs, and hence helpful in enhancing the mechanical properties of resulting composite. The high performance arises due to long-length CNTs, high level of alignment, high CNT fraction, and uniform dispersion in a polymer matrix, which are obtained simultaneously. Furthermore, the size, thickness of unidirectional composites is controlled by controlling the size of rotating mandrel diameter, sheet width, and number of rotation. The weight fraction of the CNT in composite can be controlled by optimizing the dilution of the resin, high wt% can be achieved by using a very dilute solution.

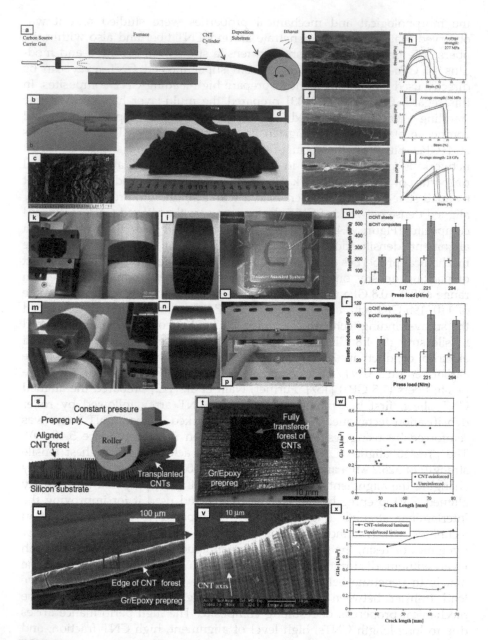

FIGURE 5.4

(a) Schematic illustration; reaction solution is sprayed into a tube reactor and pyrolysized to form a hollow CNT cylinder which is then condensed and deposited on a winding drum, (b) photograph of a hollow CNT cylinder being blown out from the reactor, (c) small piece cut from the film, (d) whole film removed from the substrate, (e, h) unaligned sample, (f, i) slightly aligned sample, (g, j) well-aligned sample [102]. (Reprinted with permission from Xu W, Chen Y, Zhan H, Wang JN. High-strength carbon nanotube film from improving alignment and densification.

FIGURE 5.4 (CAPTION CONTINUED)
Nano Letters 16, no. 2:946–952. Copyright 2016 American Chemical Society). (k) Drawing and winding technique, (l) non-pressed CNT sheet, (m) drawing, winding, and pressing process, (n) pressed CNT sheets, (o) vacuum-assisted system, (p) composite processing in hot press, (q, r) effects of pressing on the mechanical properties of aligned CNT sheets and composites [122]. (Reproduced from Mechanical property enhancement of aligned multi-walled carbon nanotube sheets and composites through press-drawing process, *Advanced Composite Materials*. 2016;25(1):73–86), (s) illustration of the "transfer-printing" process, (t) CNT forest fully transplanted from its original silicon substrate to the surface of a Gr/Epprepreg ply, (u, v) SEM images of the CNT forest, showing CNT alignment after transplantation. Example R-curves for baseline and CNT-reinforced prepreg laminates: (w) Mode I IM7/977-3; and (x) Mode II AS4/8552 [123]. (Reprinted from *Composites Part A: Applied Science and Manufacturing* 39, no. 6, Garcia EJ, Wardle BL, Hart AJ. Joining prepreg composite interfaces with aligned carbon nanotubes, 1065–1070. Copyright 2008, with permission from Elsevier).

One of the major issue that does not allow the full utilization of the reinforcing potential of CNTs is their waviness (can waviness be classified as a defect), which suggests that waviness does not carry load uniformly, cannot pack densely, and has poor intertube connection 138]. All of these factors adversely affect strength and transportation properties. Therefore, another step is added before spray winding, that is, stretching. The modified process is called stretch winding, which uses a pair of stretching rods to apply tension in CNT sheet before wrapping it over rotating mandrel [139]. With the stretching of 12%, aligned CNT/bismaleimide composites are prepared which contain 46 wt% CNT. The resulted composite successfully achieved the tensile strength of 3.8 GPa and Young's modulus of 293 MPa having the density 1.25 g/cm^3, which is lower than unidirectional carbon fiber-reinforced composite (1.6 g/cm^3). As a result, the specific tensile strength with 12% stretching ratio has reached 30% higher than best carbon fiber-reinforced composite. Apart from stretching winding technique, post stretching of prepreg also showed effective results in improving the mechanical and electrical properties [140].

The exceptional mechanical and physical properties of CNTs are the major motivation behind the research and development of high-performance composites. So far the modulus of the CNT composite of 350 GPa was achieved through mechanical alignment, functionalization followed by infiltration of bucky paper with BMI matrix [141]. When assembling CNTs into macroscopic composites, various factors are taken into account including alignment waviness, volume fraction, interaction with polymer matrix.

CNT alignment refers to the mean nanotube orientation that is parallel to the loading direction, whereas CNT waviness is a local misalignment. There are several studies related to CNT alignment and their effect on mechanical and transportation properties. In previously mentioned techniques, synergistic effect of alignment (stretching) (is the stretching nonrecoverable) and densification (spraying) is used to obtain the horizontally aligned CNT sheet. The reported mechanical, electrical, and thermal properties of aligned

CNT bucky paper-reinforced composites are superior to those of their counterparts fabricated by CNT suspension filtration technique.

5.6 Conclusions and Future Scope

Nowadays, several advanced products are developed by reinforcing the well-dispersed network of CNT into different types of polymer matrices depending on required application. These composites are prepared to exploit the promising properties of CNTs from the nanoscale to macroscopic level. However, most critical limitations that arise in the field of composite science are poor dispersion at higher loading, weak interaction between CNT and surrounding polymer, lack of orientation of CNTs in the loading direction, which unexpectedly reduce the performance of composite. Therefore, macroscopic network of CNT like CNT fiber, bucky paper are used as reinforcement for high-loading composite, which ultimately improves the multifunctional properties of composites. But, still the problem remains the same because properties are not reached to the extent of individual CNT.

Consequently, the alignment of CNT in polymer composite has been proposed to realize the effect of CNT reinforcement on a large scale. Significant research has been conducted to prepare the aligned CNT bucky paper and to develop their reinforced composites for high-end applications. Encouragingly in the past few years, various alignment techniques have been developed such as spinning of vertically aligned CNT, direct spinning from aerogel, wet spinning of CNT suspension, and mechanical pressing of aligned CNT array. These macroscale assemblies enlighten the utilization of properties of individual CNT because they possess excellent mechanical and transportation properties in the alignment direction. Though fabrication of these assemblies involves optimization of various complex parameters, their handling in many advanced applications is easy. They can be used as an alternative material in various potential applications including high-performance advanced composites in aerospace and automobile, flexible electronics (Li-ion batteries, EMI shielding, supercapacitors, etc.), anticorrosion coating, antistatic discharge coating, miniature circuits, filters, and sensors.

In spite of the progresses in the fields of CNT-related material, substantial challenges remain, especially the large-scale continuous production of reproducible oriented CNT sheet, low yield, and high cost. Advanced technique to fabricate aligned CNT assemblies on continuous bases, removing the minuscule waviness of horizontally aligned CNTs without hindering the graphitic structure, improving the interfacial properties with polymer are some major challenges that limit their use for replacing the commercially available fibers. Nevertheless, in our point of view, recent development represents that macroscopic CNT network-reinforced composite offers a unique combination of mechanical and transportation properties.

References

1. Iijima S. Helical microtubules of graphitic carbon. *Nature*. 1991;354(6348):56.
2. Treacy MJ, Ebbesen T, Gibson J. Exceptionally high Young's modulus observed for individual carbon nanotubes. *Nature*. 1996;381(6584):678.
3. Wong EW, Sheehan PE, Lieber CM. Nanobeam mechanics: elasticity, strength, and toughness of nanorods and nanotubes. *Science*. 1997;277(5334):1971–1975.
4. Coleman JN, Khan U, Blau WJ, Gun'ko YK. Small but strong: a review of the mechanical properties of carbon nanotube–polymer composites. *Carbon*. 2006;44(9):1624–1652.
5. Baughman RH, Zakhidov AA, De Heer WA. Carbon nanotubes–the route toward applications. *Science*. 2002;297(5582):787–792.
6. Kim P, Shi L, Majumdar A, McEuen PL. Thermal transport measurements of individual multiwalled nanotubes. *Physical Review Letters*. 2001;87(21):215502.
7. Demczyk BG, Wang YM, Cumings J, Hetman M, Han W, Zettl A, et al. Direct mechanical measurement of the tensile strength and elastic modulus of multiwalled carbon nanotubes. *Materials Science and Engineering: A*. 2002;334 (1–2):173–178.
8. Ebbesen T, Lezec H, Hiura H, Bennett J, Ghaemi H, Thio T. Electrical conductivity of individual carbon nanotubes. *Nature*. 1996;382(6586):54.
9. Datsyuk V, Kalyva M, Papagelis K, Parthenios J, Tasis D, Siokou A, et al. Chemical oxidation of multiwalled carbon nanotubes. *Carbon*. 2008;46 (6):833–840.
10. Avilés F, Cauich-Rodríguez J, Moo-Tah L, May-Pat A, Vargas-Coronado R. Evaluation of mild acid oxidation treatments for MWCNT functionalization. *Carbon*. 2009;47(13):2970–2975.
11. Ma P-C, Siddiqui NA, Marom G, Kim J-K. Dispersion and functionalization of carbon nanotubes for polymer-based nanocomposites: a review. *Composites Part A: Applied Science and Manufacturing*. 2010;41(10):1345–1367.
12. Garg A, Sinnott SB. Effect of chemical functionalization on the mechanical properties of carbon nanotubes. *Chemical Physics Letters*. 1998;295(4):273–278.
13. Park C, Ounaies Z, Watson KA, Crooks RE, Smith J, Jr, Lowther SE, et al. Dispersion of single wall carbon nanotubes by in situ polymerization under sonication. *Chemical Physics Letters*. 2002;364(3–4):303–308.
14. Jiang X, Bin Y, Matsuo M. Electrical and mechanical properties of polyimide–carbon nanotubes composites fabricated by in situ polymerization. *Polymer*. 2005;46(18):7418–7424.
15. Gong X, Liu J, Baskaran S, Voise RD, Young JS. Surfactant-assisted processing of carbon nanotube/polymer composites. *Chemistry of Materials*. 2000;12 (4):1049–1052.
16. Rausch J, Zhuang R-C, Mäder E. Surfactant assisted dispersion of functionalized multi-walled carbon nanotubes in aqueous media. *Composites Part A: Applied Science and Manufacturing*. 2010;41(9):1038–1046.

17. Zhu J, Kim J, Peng H, Margrave JL, Khabashesku VN, Barrera EV. Improving the dispersion and integration of single-walled carbon nanotubes in epoxy composites through functionalization. *Nano Letters.* 2003;3(8):1107–1113.

18. Wang Z, Liang Z, Wang B, Zhang C, Kramer L. Processing and property investigation of single-walled carbon nanotube (SWNT) buckypaper/epoxy resin matrix nanocomposites. *Composites Part A: Applied Science and Manufacturing.* 2004;35(10):1225–1232.

19. Gou J. Single-walled nanotube bucky paper and nanocomposite. *Polymer International.* 2006;55(11):1283–1288.

20. Wardle BL, Saito DS, García EJ, Hart AJ, de Villoria RG, Verploegen EA. Fabrication and characterization of ultrahigh-volume-fraction aligned carbon nanotube–polymer composites. *Advanced Materials.* 2008;20(14):2707–2714.

21. Olek M, Ostrander J, Jurga S, Möhwald H, Kotov N, Kempa K, et al. Layer-by-layer assembled composites from multiwall carbon nanotubes with different morphologies. *Nano Letters.* 2004;4(10):1889–1895.

22. Pereira CM, Nóvoa P, Martins M, Forero S, Hepp F, Pambaguian L. Characterization of carbon nanotube 3D-structures infused with low viscosity epoxy resin system. *Composite Structures.* 2010;92(9):2252–2257.

23. Bradford PD, Wang X, Zhao H, Maria J-P, Jia Q, Zhu Y. A novel approach to fabricate high volume fraction nanocomposites with long aligned carbon nanotubes. *Composites Science and Technology.* 2010;70(13):1980–1985.

24. Chen I-WP, Liang Z, Wang B, Zhang C. Charge-induced asymmetrical displacement of an aligned carbon nanotube buckypaper actuator. *Carbon.* 2010;48(4):1064–1069.

25. He S, Wei J, Guo F, Xu R, Li C, Cui X, et al. A large area, flexible polyaniline/buckypaper composite with a core–shell structure for efficient supercapacitors. *Journal of Materials Chemistry A.* 2014;2(16):5898–5902.

26. Chen H, Di J, Jin Y, Chen M, Tian J, Li Q. Active carbon wrapped carbon nanotube buckypaper for the electrode of electrochemical supercapacitors. *Journal of Power Sources.* 2013;237:325–331.

27. Roy S, Bajpai R, Soin N, Bajpai P, Hazra KS, Kulshrestha N, et al. Enhanced field emission and improved supercapacitor obtained from plasma-modified bucky paper. *Small.* 2011;7(5):688–693.

28. Knapp W, Schleussner D. Field-emission characteristics of carbon buckypaper. *Journal of Vacuum Science & Technology B: Microelectronics and Nanometer Structures Processing, Measurement, and Phenomena.* 2003;21(1):557–561.

29. Sharma S, Singh BP, Chauhan SS, Jyoti J, Arya AK, Dhakate S, et al. Enhanced thermomechanical and electrical properties of multiwalled carbon nanotube paper reinforced epoxy laminar composites. *Composites Part A: Applied Science and Manufacturing.* 2018;104:129–138.

30. Teotia S, Singh BP, Elizabeth I, Singh VN, Ravikumar R, Singh AP, et al. Multifunctional, robust, light-weight, free-standing MWCNT/phenolic composite paper as anodes for lithium ion batteries and EMI shielding material. *RSC Advances.* 2014;4(63):33168–33174.

31. Elizabeth I, Mathur R, Maheshwari P, Singh B, Gopukumar S. Development of SnO₂/multiwalled carbon nanotube paper as free standing anode for lithium ion batteries (LIB). *Electrochimica Acta.* 2015;176:735–742.

32. Elizabeth I, Singh BP, Bijoy TK, Reddy VR, Karthikeyan G, Singh VN, et al. In-situ conversion of multiwalled carbon nanotubes to graphene nanosheets: an

increasing capacity anode for Li Ion batteries. *Electrochimica Acta.* 2017;231:255–263.

33. Cassell AM, Raymakers JA, Kong J, Dai H. Large scale CVD synthesis of single-walled carbon nanotubes. *The Journal of Physical Chemistry B.* 1999;103 (31):6484–6492.

34. Mathur RB, Singh BP, Pande S. *Carbon nanomaterials: synthesis, structure, properties and applications.* . CRC Press, Taylor & Francis Group; ISBN: 9781498702102, 2016.

35. Wu Z, Chen Z, Du X, Logan JM, Sippel J, Nikolou M, et al. Transparent, conductive carbon nanotube films. *Science.* 2004;305(5688):1273–1276.

36. Zhang N, Yang F, Shen C, Castro J, Lee LJ. Particle erosion on carbon nanofiber paper coated carbon fiber/epoxy composites. *Composites Part B: Engineering.* 2013;54:209–214.

37. Meng Q, Manas-Zloczower I. Carbon nanotubes enhanced cellulose nanocrystals films with tailorable electrical conductivity. *Composites Science and Technology.* 2015;120:1–8.

38. Kim KS, Jakubinek MB, Martinez-Rubi Y, Ashrafi B, Guan J, O'neill K, et al. Polymer nanocomposites from free-standing, macroscopic boron nitride nanotube assemblies. *RSC Advances.* 2015;5(51):41186–41192.

39. Ouyang X, Huang W, Cabrera E, Castro J, Lee LJ. Graphene-graphene oxide-graphene hybrid nanopapers with superior mechanical, gas barrier and electrical properties. *Aip Advances.* 2015;5(1):017135.

40. Rinzler A, Liu J, Dai H, Nikolaev P, Huffman C, Rodriguez-Macias F, et al. Large-scale purification of single-wall carbon nanotubes: process, product, and characterization. *Applied Physics A: Materials Science & Processing.* 1998;67 (1):29–37.

41. Zhang J, Jiang D, Peng H-X. A pressurized filtration technique for fabricating carbon nanotube buckypaper: structure, mechanical and conductive properties. *Microporous and Mesoporous Materials.* 2014;184:127–133.

42. Jiang L, Gao L, Sun J. Production of aqueous colloidal dispersions of carbon nanotubes. *Journal of Colloid and Interface Science.* 2003;260(1):89–94.

43. Sharma S, Singh BP, Babal AS, Teotia S, Jyoti J, Dhakate S. Structural and mechanical properties of free-standing multiwalled carbon nanotube paper prepared by an aqueous mediated process. *Journal of Materials Science.* 2017;52 (12):7503–7515.

44. Liang Z, Wang B, Zhang C, Ugarte JT, Lin C-Y, Thagard J. Method for continuous fabrication of carbon nanotube networks or membrane materials. Google Patents; 2008.

45. Zhang X. Hydroentangling: a novel approach to high-speed fabrication of carbon nanotube membranes. *Advanced Materials.* 2008;20(21):4140–4144.

46. Shimoni A, Azoubel S, Magdassi S. Inkjet printing of flexible high-performance carbon nanotube transparent conductive films by "coffee ring effect". *Nanoscale.* 2014;6(19):11084–11089.

47. Zhang X, Sreekumar T, Liu T, Kumar S. Properties and structure of nitric acid oxidized single wall carbon nanotube films. *The Journal of Physical Chemistry B.* 2004;108(42):16435–16440.

48. Pöhls J-H, Johnson MB, White MA, Malik R, Ruff B, Jayasinghe C, et al. Physical properties of carbon nanotube sheets drawn from nanotube arrays. *Carbon.* 2012;50(11):4175–4183.

49. Kuznetsov AA, Fonseca AF, Baughman RH, Zakhidov AA. Structural model for dry-drawing of sheets and yarns from carbon nanotube forests. *ACS Nano.* 2011;5(2):985–993.

50. Jung Y, Song J, Huh W, Cho D, Jeong Y. Controlling the crystalline quality of carbon nanotubes with processing parameters from chemical vapor deposition synthesis. *Chemical Engineering Journal.* 2013;228:1050–1056.

51. Stallard J, Tan W, Smail FR, Gspann T, Boies AM, Fleck NA. The mechanical and electrical properties of direct-spun carbon nanotube mats. *Extreme Mechanics Letters.* 2018;21:65–75.

52. Liu K, Sun Y, Zhou R, Zhu H, Wang J, Liu L, et al. Carbon nanotube yarns with high tensile strength made by a twisting and shrinking method. *Nanotechnology.* 2009;21(4):045708.

53. Stahl JJ, Bogdanovich AE, Bradford PD. Carbon nanotube shear-pressed sheet interleaves for mode I interlaminar fracture toughness enhancement. *Composites Part A: Applied Science and Manufacturing.* 2016;80:127–137.

54. Liu K, Sun Y, Chen L, Feng C, Feng X, Jiang K, et al. Controlled growth of super-aligned carbon nanotube arrays for spinning continuous unidirectional sheets with tunable physical properties. *Nano Letters.* 2008;8(2):700–705.

55. Jiang K, Li Q, Fan S. Nanotechnology: spinning continuous carbon nanotube yarns. *Nature.* 2002;419(6909):801.

56. De Volder MF, Tawfick SH, Baughman RH, Hart AJ. Carbon nanotubes: present and future commercial applications. *Science.* 2013;339(6119):535–539.

57. Tran C-D, Humphries W, Smith SM, Huynh C, Lucas S. Improving the tensile strength of carbon nanotube spun yarns using a modified spinning process. *Carbon.* 2009;47(11):2662–2670.

58. Park J, Lee K-H. Carbon nanotube yarns. *Korean Journal of Chemical Engineering.* 2012;29(3):277–287.

59. Zhao H, Zhang Y, Bradford PD, Zhou Q, Jia Q, Yuan F-G, et al. Carbon nanotube yarn strain sensors. *Nanotechnology.* 2010;21(30):305502.

60. Zhang M, Fang S, Zakhidov AA, Lee SB, Aliev AE, Williams CD, et al. Strong, transparent, multifunctional, carbon nanotube sheets. *Science.* 2005;309 (5738):1215–1219.

61. Zhang Q, Zhao M-Q, Huang J-Q, Liu Y, Wang Y, Qian W-Z, et al. Vertically aligned carbon nanotube arrays grown on a lamellar catalyst by fluidized bed catalytic chemical vapor deposition. *Carbon.* 2009;47(11):2600–2610.

62. Lepro X, Lima MD, Baughman RH. Spinnable carbon nanotube forests grown on thin, flexible metallic substrates. *Carbon.* 2010;48(12):3621–3627.

63. Zhu H, Xu C, Wu D, Wei B, Vajtai R, Ajayan P. Direct synthesis of long single-walled carbon nanotube strands. *Science.* 2002;296(5569):884–886.

64. Li Y-L, Kinloch IA, Windle AH. Direct spinning of carbon nanotube fibers from chemical vapor deposition synthesis. *Science.* 2004;304(5668):276–278.

65. Koziol K, Vilatela J, Moisala A, Motta M, Cunniff P, Sennett M, et al. High-performance carbon nanotube fiber. *Science.* 2007;318(5858):1892–1895.

66. Zhong XH, Li YL, Liu YK, Qiao XH, Feng Y, Liang J, et al. Continuous multilayered carbon nanotube yarns. *Advanced Materials.* 2010;22(6):692–696.

67. Motta M, Moisala A, Kinloch IA, Windle AH. High performance fibres from "dog bone" carbon nanotubes. *Advanced Materials.* 2007;19(21):3721–3726.

68. Yang HH. *Aromatic high-strength fibers.* Wiley, USA; 1989.

69. Vigolo B, Penicaud A, Coulon C, Sauder C, Pailler R, Journet C, et al. Macroscopic fibers and ribbons of oriented carbon nanotubes. *Science*. 2000;290 (5495):1331–1334.
70. Ericson LM, Fan H, Peng H, Davis VA, Zhou W, Sulpizio J, et al. Macroscopic, neat, single-walled carbon nanotube fibers. *Science*. 2004;305(5689):1447–1450.
71. Ramesh S, Ericson LM, Davis VA, Saini RK, Kittrell C, Pasquali M, et al. Dissolution of pristine single walled carbon nanotubes in superacids by direct protonation. *The Journal of Physical Chemistry B*. 2004;108(26):8794–8798.
72. Parra-Vasquez ANG, Behabtu N, Green MJ, Pint CL, Young CC, Schmidt J, et al. Spontaneous dissolution of ultralong single-and multiwalled carbon nanotubes. *ACS Nano*. 2010;4(7):3969–3978.
73. Krupke R, Hennrich F, Weber H, Kappes M, Löhneysen HV. Simultaneous deposition of metallic bundles of single-walled carbon nanotubes using ac-dielectrophoresis. *Nano Letters*. 2003;3(8):1019–1023.
74. Bornhoeft LR, Castillo AC, Smalley PR, Kittrell C, James DK, Brinson BE, et al. Teslaphoresis of carbon nanotubes. *ACS Nano*. 2016;10(4):4873–4881.
75. Davis VA, Parra-Vasquez ANG, Green MJ, Rai PK, Behabtu N, Prieto V, et al. True solutions of single-walled carbon nanotubes for assembly into macroscopic materials. *Nature Nanotechnology*. 2009;4(12):830.
76. Wang D, Song P, Liu C, Wu W, Fan S. Highly oriented carbon nanotube papers made of aligned carbon nanotubes. *Nanotechnology*. 2008;19(7):075609.
77. Pint CL, Xu Y-Q, Pasquali M, Hauge RH. Formation of highly dense aligned ribbons and transparent films of single-walled carbon nanotubes directly from carpets. *ACS Nano*. 2008;2(9):1871–1878.
78. Hayamizu Y, Yamada T, Mizuno K, Davis RC, Futaba DN, Yumura M, et al. Integrated three-dimensional microelectromechanical devices from processable carbon nanotube wafers. *Nature Nanotechnology*. 2008;3(5):289.
79. Rashid MH-O, Pham SQ, Sweetman LJ, Alcock LJ, Wise A, Nghiem LD, et al. Synthesis, properties, water and solute permeability of MWNT buckypapers. *Journal of Membrane Science*. 2014;456:175–184.
80. Zhang L, Zhang G, Liu C, Fan S. High-density carbon nanotube buckypapers with superior transport and mechanical properties. *Nano Letters*. 2012;12 (9):4848–4852.
81. Pham GT, Park Y-B, Wang S, Liang Z, Wang B, Zhang C, et al. Mechanical and electrical properties of polycarbonate nanotube buckypaper composite sheets. *Nanotechnology*. 2008;19(32):325705.
82. Han J-H, Zhang H, Chen M-J, Wang G-R, Zhang Z. CNT buckypaper/ thermoplastic polyurethane composites with enhanced stiffness, strength and toughness. *Composites Science and Technology*. 2014;103:63–71.
83. Qu B, Zhuo D, Wang R, Wu L, Cheng X. Enhancement of mechanical properties of buckypapers/polyethylene composites by microwave irradiation. *Composites Science and Technology*. 2018;164:313–318.
84. Jakubinek MB, Ashrafi B, Guan J, Johnson MB, White MA, Simard B. 3D chemically cross-linked single-walled carbon nanotube buckypapers. *RSC Advances*. 2014;4(101):57564–57573.
85. Kobashi K, Hirabayashi T, Ata S, Yamada T, Futaba DN, Hata K Green, scalable, binderless fabrication of a single-walled carbon nanotube nonwoven fabric based on an ancient Japanese paper process. *ACS Applied Materials & Interfaces*. 2013;5(23):12602–12608.

86. Dettlaff-Weglikowska U, Skákalová V, Graupner R, Jhang SH, Kim BH, Lee HJ, et al. Effect of SOCl₂ treatment on electrical and mechanical properties of single-wall carbon nanotube networks. *Journal of the American Chemical Society*. 2005;127(14):5125–5131.

87. Park JG, Smithyman J, Lin C-Y, Cooke A, Kismarahardja AW, Li S, et al. Effects of surfactants and alignment on the physical properties of single-walled carbon nanotube buckypaper. *Journal of Applied Physics*. 2009;106(10):104310.

88. Zhang J, Jiang D, Peng H-X QF. Enhanced mechanical and electrical properties of carbon nanotube buckypaper by in situ cross-linking. *Carbon*. 2013;63:125–132.

89. Arif MF, Kumar S, Shah T. Tunable morphology and its influence on electrical, thermal and mechanical properties of carbon nanostructure-buckypaper. *Materials & Design*. 2016;101:236–244.

90. Spitalsky Z, Tsoukleri G, Tasis D, Krontiras C, Georga S, Galiotis C. High volume fraction carbon nanotube–epoxy composites. *Nanotechnology*. 2009;20 (40):405702.

91. Che J, Chen P, Chan-Park MB. High-strength carbon nanotube buckypaper composites as applied to free-standing electrodes for supercapacitors. *Journal of Materials Chemistry A*. 2013;1(12):4057–4066.

92. Liu Q, Li M, Wang Z, Gu Y, Li Y, Zhang Z. Improvement on the tensile performance of buckypaper using a novel dispersant and functionalized carbon nanotubes. *Composites Part A: Applied Science and Manufacturing*. 2013;55:102–109.

93. Byrne MT, Hanley CA, Gun'ko YK. Preparation and properties of buckypaper–gold nanoparticle composites. *Journal of Materials Chemistry*. 2010;20 (15):2949–2951.

94. Whitten PG, Spinks GM, Wallace GG. Mechanical properties of carbon nanotube paper in ionic liquid and aqueous electrolytes. *Carbon*. 2005;43(9):1891–1896.

95. Zhang L, Wang X, Xu W, Zhang Y, Li Q, Bradford PD, et al. Strong and conductive dry carbon nanotube films by microcombing. *Small*. 2015;11 (31):3830–3836.

96. Malik R, McConnell C, Alvarez NT, Haase M, Gbordzoe S, Shanov V. Rapid, in situ plasma functionalization of carbon nanotubes for improved CNT/epoxy composites. *RSC Advances*. 2016;6(110):108840–108850.

97. Chaffee J, Lashmore D, Lewis D, Mann J, Schauer M, White B, et al. Direct synthesis of CNT yarns and sheets. *Nsti Nanotech 2008*. 2008;3:118–121.

98. Chen I-WP. Noncovalently functionalized highly conducting carbon nanotube films with enhanced doping stability via an amide linkage. *Chemical Communications*. 2013;49(27):2753–2755.

99. Xu F, Wei B, Liu W, Zhu H, Zhang Y, Qiu Y. In-plane mechanical properties of carbon nanotube films fabricated by floating catalyst chemical vapor decomposition. *Journal of Materials Science*. 2015;50(24):8166–8174.

100. Liu P, Tan YF, Hu DC, Jewell D, Duong HM. Multi-property enhancement of aligned carbon nanotube thin films from floating catalyst method. *Materials & Design*. 2016;108:754–760.

101. Zhang Z, Gu Y, Wang S, Li Q, Li M, Zhang Z. Enhanced dielectric and mechanical properties in chlorine-doped continuous CNT sheet reinforced sandwich polyvinylidene fluoride film. *Carbon*. 2016;107:405–414.

102. Xu W, Chen Y, Zhan H, Wang JN. High-strength carbon nanotube film from improving alignment and densification. *Nano Letters*. 2016;16(2):946–952.

103. Han B, Xue X, Xu Y, Zhao Z, Guo E, Liu C, et al. Preparation of carbon nanotube film with high alignment and elevated density. *Carbon.* 2017;122:496–503.

104. Zhang M, Li M, Wang S, Wang Y, Zhang Y, Gu Y, et al. The loading-rate dependent tensile behavior of CNT film and its bismaleimide composite film. *Materials & Design.* 2017;117:37–46.

105. Zhang S, Park JG, Nguyen N, Jolowsky C, Hao A, Liang R. Ultra-high conductivity and metallic conduction mechanism of scale-up continuous carbon nanotube sheets by mechanical stretching and stable chemical doping. *Carbon.* 2017;125:649–658.

106. Jiang Q, Li Y, Xie J, Sun J, Hui D, Qiu Y. Plasma functionalization of bucky paper and its composite with phenylethynyl-terminated polyimide. *Composites Part B: Engineering.* 2013;45(1):1275–1281.

107. Sahoo NG, Rana S, Cho JW, Li L, Chan SH. Polymer nanocomposites based on functionalized carbon nanotubes. *Progress in Polymer Science.* 2010;35 (7):837–867.

108. Balasubramanian K, Burghard M. Chemically functionalized carbon nanotubes. *Small.* 2005;1(2):180–192.

109. An KH, Heo JG, Jeon KG, Bae DJ, Jo C, Yang CW, et al. X-ray photoemission spectroscopy study of fluorinated single-walled carbon nanotubes. *Applied Physics Letters.* 2002;80(22):4235–4237.

110. Velasco-Soto M, Pérez-García S, Rychwalski R, Licea-Jiménez L. Dispersion of carbon nanomaterials. In *Nanocolloids: a meeting point for scientists and technologists.* edited by Margarita Sanchez-Dominguez, Carlos Rodriguez-Abreu, Cambridge USA Elsevier; 2016. p. 247.

111. Wang S, Liang Z, Wang B, Zhang C. High-strength and multifunctional macroscopic fabric of single-walled carbon nanotubes. *Advanced Materials.* 2007;19(9):1257–1261.

112. Vohrer U, Zschoerper NP, Koehne Y, Langowski S, Oehr C. Plasma modification of carbon nanotubes and bucky papers. *Plasma Processes and Polymers.* 2007;4(S1):S871–S877.

113. Naseh MV, Khodadadi AA, Mortazavi Y, Pourfayaz F, Alizadeh O, Maghrebi M. Fast and clean functionalization of carbon nanotubes by dielectric barrier discharge plasma in air compared to acid treatment. *Carbon.* 2010;48(5):1369–1379.

114. Iijima S, Brabec C, Maiti A, Bernholc J. Structural flexibility of carbon nanotubes. *The Journal of Chemical Physics.* 1996;104(5):2089–2092.

115. Winey KI, Kashiwagi T, Mu M. Improving electrical conductivity and thermal properties of polymers by the addition of carbon nanotubes as fillers. *MRS Bulletin.* 2007;32(4):348–353.

116. Pande S, Chaudhary A, Patel D, Singh BP, Mathur RB. Mechanical and electrical properties of multiwall carbon nanotube/polycarbonate composites for electrostatic discharge and electromagnetic interference shielding applications. *RSC Advances.* 2014;4(27):13839–13849.

117. Saini P, Choudhary V, Singh B, Mathur R, Dhawan S. Polyaniline–MWCNT nanocomposites for microwave absorption and EMI shielding. *Materials Chemistry and Physics.* 2009;113(2–3):919–926.

118. Babal AS, Gupta R, Singh BP, Singh VN, Dhakate SR, Mathur RB. Mechanical and electrical properties of high performance MWCNT/polycarbonate composites

    prepared by an industrial viable twin screw extruder with back flow channel. *RSC Advances*. 2014;4(110):64649–64658.

119. Martin C, Sandler J, Windle A, Schwarz M-K, Bauhofer W, Schulte K, et al. Electric field-induced aligned multi-wall carbon nanotube networks in epoxy composites. *Polymer*. 2005;46(3):877–886.

120. Steinert BW, Dean DR. Magnetic field alignment and electrical properties of solution cast PET–carbon nanotube composite films. *Polymer*. 2009;50 (3):898–904.

121. Xie X-L, Mai Y-W, Zhou X-P. Dispersion and alignment of carbon nanotubes in polymer matrix: a review. *Materials Science and Engineering: R: Reports*. 2005;49 (4):89–112.

122. Nam TH, Goto K, Oshima K, Premalal E, Shimamura Y, Inoue Y, et al. Mechanical property enhancement of aligned multi-walled carbon nanotube sheets and composites through press-drawing process. *Advanced Composite Materials*. 2016;25(1):73–86.

123. Garcia EJ, Wardle BL, Hart AJ. Joining prepreg composite interfaces with aligned carbon nanotubes. *Composites Part A: Applied Science and Manufacturing*. 2008;39(6):1065–1070.

124. Sharma S, Kumar V, Pathak AK, Yokozeki T, Yadav SK, Singh VN, et al. Design of MWCNT bucky paper reinforced PANI–DBSA–DVB composites with superior electrical and mechanical properties. *Journal of Materials Chemistry C*. 2018;6(45):12396–12406.

125. Kumar V, Sharma S, Pathak A, Singh BP, Dhakate SR, Yokozeki T, et al. Interleaved MWCNT buckypaper between CFRP laminates to improve through-thickness electrical conductivity and reducing lightning strike damage. *Composite Structures*. 2019;210:581–589.

126. Zhang M, Atkinson KR, Baughman RH. Multifunctional carbon nanotube yarns by downsizing an ancient technology. *Science*. 2004;306(5700):1358–1361.

127. Motta M, Li Y-L, Kinloch I, Windle A. Mechanical properties of continuously spun fibers of carbon nanotubes. *Nano Letters*. 2005;5(8):1529–1533.

128. Zhang S, Koziol KK, Kinloch IA, Windle AH. Macroscopic fibers of well-aligned carbon nanotubes by wet spinning. *Small*. 2008;4(8):1217–1222.

129. Zhang X, Li Q, Holesinger TG, Arendt PN, Huang J, Kirven PD, et al. Ultra-strong, stiff, and lightweight carbon-nanotube fibers. *Advanced Materials*. 2007;19(23):4198–4201.

130. Tran C-D, Lucas S, Phillips D, Randeniya L, Baughman R, Tran-Cong T. Manufacturing polymer/carbon nanotube composite using a novel direct process. *Nanotechnology*. 2011;22(14):145302.

131. Bogdanovich AE, Bradford PD. Carbon nanotube yarn and 3-D braid composites. Part I: tensile testing and mechanical properties analysis. *Composites Part A: Applied Science and Manufacturing*. 2010;41(2):230–237.

132. Bradford PD, Bogdanovich AE. Carbon nanotube yarn and 3-D braid composites. Part II: dynamic mechanical analysis. *Composites Part A: Applied Science and Manufacturing*. 2010;41(2):238–246.

133. Cheng Q, Wang J, Wen J, Liu C, Jiang K, Li Q, et al. Carbon nanotube/epoxy composites fabricated by resin transfer molding. *Carbon*. 2010;48(1):260–266.

134. Coleman JN, Cadek M, Blake R, Nicolosi V, Ryan KP, Belton C, et al. High performance nanotube-reinforced plastics: understanding the mechanism of strength increase. *Advanced Functional Materials*. 2004;14(8):791–798.

135. Shim BS, Zhu J, Jan E, Critchley K, Ho S, Podsiadlo P, et al. Multiparameter structural optimization of single-walled carbon nanotube composites: toward record strength, stiffness, and toughness. *ACS Nano*. 2009;3(7):1711–1722.
136. Hou Y, Tang J, Zhang H, Qian C, Feng Y, Liu J. Functionalized few-walled carbon nanotubes for mechanical reinforcement of polymeric composites. *ACS Nano*. 2009;3(5):1057–1062.
137. Liu K, Sun Y, Lin X, Zhou R, Wang J, Fan S, et al. Scratch-resistant, highly conductive, and high-strength carbon nanotube-based composite yarns. *ACS Nano*. 2010;4(10):5827–5834.
138. Zhang Y, Sheehan CJ, Zhai J, Zou G, Luo H, Xiong J, et al. Polymer-embedded carbon nanotube ribbons for stretchable conductors. *Advanced Materials*. 2010;22(28):3027–3031.
139. Wang X, Yong Z, Li Q, Bradford PD, Liu W, Tucker DS, et al. Ultrastrong, stiff and multifunctional carbon nanotube composites. *Materials Research Letters*. 2013;1(1):19–25.
140. Wang X, Bradford PD, Liu W, Zhao H, Inoue Y, Maria J-P, et al. Mechanical and electrical property improvement in CNT/Nylon composites through drawing and stretching. *Composites Science and Technology*. 2011;71(14):1677–1683.
141. Cheng Q, Wang B, Zhang C, Liang Z. Functionalized carbon-nanotube sheet/ bismaleimide nanocomposites: mechanical and electrical performance beyond carbon-fiber composites. *Small*. 2010;6(6):763–767.

6

Modeling Multiwalled Carbon Nanotubes: From Quantum Mechanical Calculations to Mechanical Analogues

Brijesh Kumar Mishra

Computational Chemistry Unit
International Institute of Information Technology
Bangalore, India

Balakrishnan Ashok

Centre for Complex Systems & Soft Matter Physics
International Institute of Information Technology
Bangalore, India

6.1 Introduction

The potential use of double-walled carbon nanotubes (DWCNTs) as nanogears, rotors, and ratchets has often been discussed in the context of nanoelectromechanical systems and nanotechnological applications. In this chapter, we discuss how one can use quantum mechanical and classical calculations to investigate favorable configurations of multi- and double-walled carbon nanotubes (MWCNTs and DWCNTs). Even more interestingly, we model the DWCNT system through a mechanical analogue of coaxial cylinders with a nonlinear spring interaction, which enables us to reproduce the interaction energy of the system obtained through quantum-mechanical density functional theory (DFT) calculations.

This also permits us to predict the elastic spring constant of the system and the natural frequency of small amplitude axial vibrations; these predictions are found to be in agreement with those available in the published literature. Such a modeling paradigm of quantum-mechanical DFT calculations used along with a classical nonlinear spring model potentially enables one to tailor the composition of the nanotubes to yield desired mechanical and dynamical behavior.

These calculations would be invaluable in designing structures and networks of CNTs and DWCNTs, and in predicting their dynamical behavior.

CNTs have mechanical, electronic, and transport properties that make them of great practical importance. With tensile strength in the order of hundreds of gigapascal and elastic moduli in several terapascal, CNTs can potentially be used to make exceptionally strong cables and tethers. Other applications include their use as quantum wires, as well as light-sensitive field effect transistors.

Before proceeding further, a very brief review of the geometry and structure of CNTs and DWCNTs is warranted in order to keep this discussion self-contained. For details, the reader is referred to any of the several reviews and texts in the published literature on the subject, for instance Dresselhaus et al.,[1] Wong and Akinwande,[2] or Rafii-Tabar.[3]

CNTs are obtained by the rolling of graphene sheets. If $\mathbf{a}_1$ and $\mathbf{a}_2$ are the primitive vectors of the unit cell, the chiral vector $\mathbf{C}_h = n\mathbf{a}_1 + m\mathbf{a}_2$ is related to the angle of twist and tube diameter. The chiral indices (n,m) define the geometry of a nanotube – whether they are zigzag $(n,0)$, chiral (n,m), or armchair (n,n). Tube diameter D of a CNT is calculated from

$$D = \frac{a_{c-c}}{\pi}\sqrt{3(n^2 + m^2 + nm)},$$ with a_{c-c} being the carbon–carbon bond length. The effects of curvature and geometry must be considered when fabricating nanotubes from graphene sheets, as the radius of the nanotube and the curvature energy are intimately connected. When two or more concentric SWCNTs are located coaxially, a DWCNT or MWCNT is formed. Examples of such structures are shown in Figure 6.1.

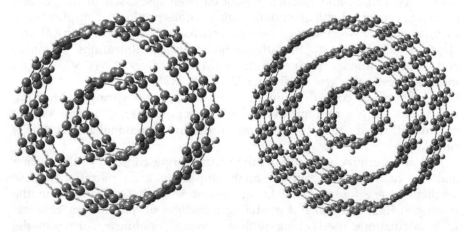

FIGURE 6.1
Schematic representations of a part of a DWCNT, at left, and a triple-walled carbon nanotube, at right.

Mutual interactions between the walls, including van der Waals interactions, help stabilize the CNT structures. If the inner tube of a DWCNT is of type (n,n) and the outer tube is of type (m,m), the structure would be denoted as being of type $(n,n)@(m,m)$.

Investigations of the behavior of DWCNTs have indicated that relative motion of the two coaxial tubes occurs with very low friction.[4] The natural idea of considering these as nanosprings has lead to several numerical and analytical studies of their oscillatory behavior in the literature.[5–7] DWCNT assemblies can also be used as nanogears, rotors, and ratchets, and have great potential for use as components in nanoelectromechanical machines.

To theoretically see how DWCNT systems behave and what stable configurations can be achieved, one can employ various methods. One method, the results of which we discuss here, is to use quantum mechanical DFT calculations to obtain the interaction energies and configurations of stable systems, and then try and model the system through an intuitive, classical system in order to understand – and predict – the kinetics and dynamics of the CNTs.

6.2 Density Functional Calculations for the CNT Systems

DFT calculations are useful when one wants to probe a system to find its material properties. There are various basis sets and optimization techniques that are often used in DFT calculations, the details of which we shall not discuss here. The interested reader is referred to any standard text or reference for more information, such as, for example, the book by Levine.[8] DFT calculations whose results were reported in Mishra and Ashok[9] and are discussed in this chapter were performed using the TURBOMOLE package.[10,11]

The geometries of the nanotubes were optimized using the dispersion-corrected DFT (BLYP-D3 functional) and the double-zeta (def2-SVP) quality basis set.[12] Interaction energies for the coaxial nanotubes were calculated using the BLYP-D3 method with triple-zeta valence polarized (def2-TZVP) basis set. Popov and coworkers[13] have showed that the DFT-D3 method gives accurate interaction energies for stacked graphene systems. All the scan calculations for slide/roll motion were performed using fixed coordinates. Hydrogen atoms were added manually to satisfy the valency of the carbon atoms at the periphery The H and C atoms are very similar in terms of their electronegativity, therefore, the C–H bonds on the edge of the tube will have a very weak dipole nature. It can therefore be assumed that the effect of such bond polarity would be negligible on the total interaction energy. The geometry of the nanogears could not be optimized due to limited computational resources, and coordinates of the atoms were fixed during calculations. Resolution of identity (RI) approximation was used to speed up the calculations.[14] For larger systems, multipole accelerated RI-J (MARI-J) approximation was also used.[15] Interaction energies (ΔE) for the DWCNTs were computed thus: $\Delta E = E(nn@mm) - (E(nn) + E(mm))$, where $E(nn)$ and $E(mm)$ are energy values for SWCNT, while $E(nn@mm)$ is the energy value for the DWCNT. Similarly, the triple-walled CNT interaction energy was calculated using:

$$\Delta E = E(nn@mm@kk) - (E(nn) + E(mm) + E(kk)).$$

We visualize the DWCNT system to be but a nanoscale analogue of a macroscopic spring or damper. The inner nanotube, we suggest, would be free to move out telescopically along the axis. DFT calculations performed as indicated above are in consonance with this. The calculations were performed for DWCNTs of different lengths, both 0.6 and 1.2 nm.[9]

6.3 Modeling the DWCNT through a Mechanical Analogue: A Nanospring

The structure of CNTs immediately suggests to the mind the possibility of using them as nanoscale tubes and conduits, and as components of gears, bearings, rotors, and so on at this scale. Figure 6.2 is an illustration of a DWCNT with its inner tube partially displaced out along the axis.

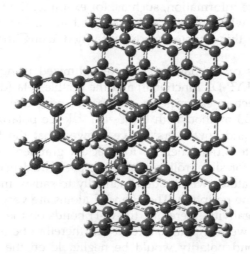

FIGURE 6.2
Side view of a DWCNT nanospring, showing the inner nanotube to have moved out, along the axis.

For a practical implementation to be possible, interactions between coaxial carbon tubes must be quantifiable and predictable. One way of doing this is by modeling it as a simpler, analogous mechanical system by considering the geometry of the CNTs, as reported by Mishra and Ashok,[9] and which we discuss here. A DWCNT system can therefore be thought of as two concentric cylinders, with some interaction potentials determining their behavior. That there can be a mechanical interaction between the two walls of a DWCNT has been indicated earlier.[16,17] The walls of the concentric cylinders separated by a distance d are modeled to have a short-range $1/d^6$ van der Waals interaction, and can be visualized as concentric rings, each ring of length L_m, which are placed one atop the other to form a DWCNT of length L. In addition, a mechanical coupling in the form of nonlinear spring acts between the two concentric CNTs, limited by the axial distance z to which the inner cylinder moves out of the outer one. If there are N atoms on the inner rings, elastic and harmonic interaction is modeled through N springs acting radially between the two CNT walls. Spring softening occurs as the inner CNT moves out of the outer tube, a distance z along the central axis, and each of the elastic units has a limited influence to a length no larger than L/L_m.

An assumption of uniform distribution of the N springs along the circumference of the inner CNT allows one to resolve the restoring elastic force into its radial components and the component parallel to the axis. Apart from the non-Hookean spring and the van der Waals interaction mentioned above, a radial harmonic potential acts between the two CNTs, not affecting that segment of the tube that has moved out. Keeping these arguments in mind, we are able to obtain an expression for the interaction energy E of a DWCNT of length L whose inner tube has axially extended by a distance z. Because the elastic spring force is limited by the distance moved by the inner tube, we get distinct expressions for the interaction energy depending upon the location of the inner CNT. For smaller extensions, the interaction energy E_1 (found by integrating the effective restoring force on the inner tube and including the van der Waals and radial harmonic terms) is given by[9]

$$E_1 = E_0 + \frac{Nkz^2}{2} - \frac{Nkz^3}{3L} - \frac{Nkd^3}{L}\tan^{-1}\left(\frac{z}{d}\right) + \frac{(pNkd^2)(L-Z)}{2L}$$
$$- Nkd^2\left(\sec\left(\tan^{-1}(z/d)\right) - 1\right) + \frac{c}{d^6},$$

while for tube displacements longer than the length of a molecular unit, the interaction energy E_2 is given by

$$E_2 = E'_o - \frac{Nkz^3}{3L} + \frac{c}{d^6} + \frac{(pNkd^2)(L-Z)}{2L}$$

In the equations above, E_0 and E_0' are constants of integration, while p can be termed as a tuning parameter for the harmonic interaction; this interaction is modeled as occurring between the overlapping walls of the CNTs and reduces to zero when the inner tube's translational displacement takes it completely outside, when $z = L$. We would therefore expect this to also have a functional dependence on the number of atoms present on a ring in the inner tube, N, and find from the data that p scales as $p \sim N^{-0.32}$ (see Mishra and Ashok[9]). From these results, it is clear that when the inner CNT is completely inside, and no axial translational displacement has occurred, the interaction energy will be dependent only on the interwall separation d.

Figure 6.3 illustrates the dependence of interaction energy of DWCNTs on axial displacement. The data points shown are for four different stable armchair–armchair DWCNT systems, of type $(n,n)@(n+5,n+5)$, with different number of atoms N on the inner ring, such that $N4 < N3 < N2 < N1$, but each DWCNT of the same length. The data points have been obtained from DFT calculations. The solid curves are illustrative of the values obtained from the mechanical model, $E1$ and $E2$ are corresponding interaction energies for the system with the inner tube having undergone small and large axial displacements, respectively.

Obtaining numerical values for spring constant k from the data by fitting the mechanical model then enables us to predict the natural frequency of oscillation for small amplitude axial displacements of the inner CNT out of

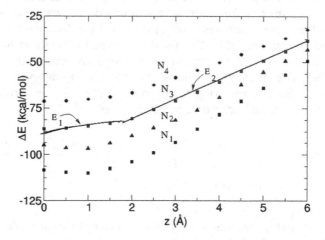

FIGURE 6.3

Interaction energies of DWCNTs plotted as a function of axial displacement, data points obtained from DFT calculations for four different configurations, with different number of atoms on a ring in the inner CNT, $N4 < N3 < N2 < N1$. Solid curves are illustrative of curves obtained from the mechanical model, delineating distinct behaviors for small translations ($E1$) and larger ones ($E2$). For details, please see Mishra and Ashok.[9]

the DWCNT. The natural frequency of oscillation is obtained by considering the system in the linear limit and considering the usual relation between frequency and the spring constant in the Hookean limit. For a nonlinear system, one hallmark is an amplitude-dependent oscillatory frequency. The oscillatory frequencies (depending upon large nanotube displacements) would be obtainable once we consider dynamics in the system, and look at time-dependent behavior. Spring constant values are found to be in the range of approximately 0.34 to 0.41 N/m (for DWCNTs that are 0.6 and 1.2 nm long, respectively). These translate to natural frequencies of oscillation $\nu \approx 78 - 109 GHz$ for 0.6 nm DWCNTs and $\nu \approx 60 - 85 GHz$ for 1.2 nm DWCNTs. The range in frequencies occurs even for the same length DWCNT because of the range in masses of the systems, depending upon how many atoms are present.

These values are also in agreement with others reported in the literature.[6,18,19]

6.4 Nanobearings, Ratchet Wheels, and Nanogears

Thus far, we have looked at the translational motion of CNTs in a DWCNT and seen how a nice mechanical analogue can be drawn upon for the nanospring to give very nice and remarkably consistent and useful values with respect to the system's effective spring constant and natural frequency of axial vibrations.

Considering now the mutual rotation of CNTs, it is seen that a change in angular position leads to a near-negligible change in interaction energy. This change, for most cases, is as low as 0.2–0.4 kcal/mol, and for some configurations it may be approximately 1 kcal/mol, as seen by Mishra and Ashok.[9] For most cases, the interaction energy profiles seem very low and make a case for considering DWCNTs as ideal candidates for nanobearings that would have frictionless and smooth, uniform rotation as a characteristic.

Modification of DWCNTs by symmetrical addition of four acetylide groups leads to the formation of a nanoscale ratchet wheel, as shown in Figure 6.4. Directional CH-pi bond formation causes stable structures at 0°, 90° and 180°. At 45° and 135°, the system is unstable due to these bonds breaking. The directionality of the acetylide group causes a nonuniform energy profile, which enables the use of the DWCNT as a ratchet wheel. One can tune the profile according to their requirements by appropriate addition of substituents.

Nanogears may likewise be constructed out of DWCNTs by addition of substituents such as phenyl groups that would act as the gear-teeth, as illustrated in Figure 6.5. The change in the interaction energy for various angles of rotation is seen to be very small, less than 1 kcal/mol, and rotation would cause breakage of pi–pi interaction bonds and new formation of pi–pi interactions. This system too is a candidate for being considered as a near-zero-friction gear, with room-temperature thermal energy likely to be

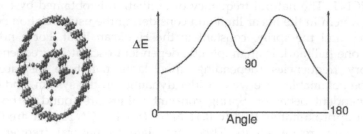

FIGURE 6.4
Ratchet wheel with acetylide groups on the four arms attached to the inner tube. Rough Sketch of the interaction energy as a function of angle for the inner tube.

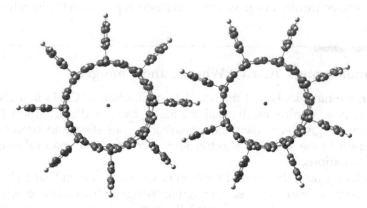

FIGURE 6.5
Addition of phenyl groups as arms to SWCNTs gives rise to frictionless nanogears.

sufficient to rotate it. DFT (Kohn–Sham equations) uses the density of the electrons, which is calculated using the approximate wave function of the system. DFT accounts for all the interactions between all the electrons and nuclei. There is an additional fitted exchange-correlation function. So the calculations treat the system as a collection of particles kept at certain distances, and inclusion of roughness would need a different approach, where one specifically includes defects in the molecular structure.

6.5 Other Structures

B–N nanotubes are structural analogues of CNTs, where carbon atoms are alternately substituted by nitrogen and boron atoms. B–N nanotubes were first synthesized by the arc discharge method. Since then, various other

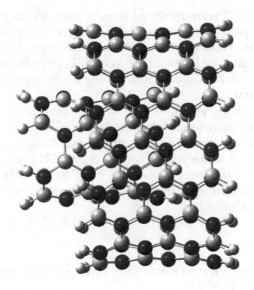

FIGURE 6.6
Double-walled B–N nanotube forming a nanospring.

experimental techniques have been developed to synthesize pure B–N nanotubes.

B–N nanotubes have an advantage over CNTs as they have higher resistance to oxidation and greater thermal stability. Structures such as nanosprings can also be formed, as illustrated in Figure 6.6.

6.6 Stability of DWCNT Structures and Their Practical Applications

Stable DWCNT structures discussed above fall in an energy range of approximately 40–120 kcal/mol. They are therefore expected to be stable even under thermal fluctuations.

Rotational actuators based on CNTs constructed by Fennimore et al.[20] had the outer diameter of the MWCNTs used varying from 10 to 40 nm. Characterization studies reported in the literature (Hirschmann et al.[21] and Kim et al.[22]) show that catalytically grown DWCNTs of the dimensions similar to the structures theoretically studied in Mishra and Ashok[9] and discussed here may be formed in the laboratory.

While a motor constructed by Fennimore et al.[20] had a 10 nm diameter bearing, Bourlon et al.[23] constructed a nanoelectromechanical system comprising a plate rotating around a multiwalled nanotube bearing. Outer layers of the MWCNT were removed therein through electrical breakdown and the rotating plate attached to the inner CNT that turned inside the (fixed) outer shells. Transmission electron microscope studies by Zettl and collaborators[24] on the core layers of the MWCNTs have focused on telescopic motion of MWCNTs, the outer layers having again been removed through electrical breakdown.

Making nanosprings, gears, ratchet wheels, and rotors is increasingly becoming more and more feasible in the laboratory. Recently, many advances have been made in theoretical and computational studies of nanomechanics as well.[25] The use of nanosprings, rotors, etc. in CNT networks as part of everyday practical machinery would be possible once their physical and dynamical properties are well understood, and once constructing them on a mass scale becomes a reality.

Acknowledgments

B.K.M. acknowledges support from SERB, DST, Government of India through Project No.ECR/2015/000274.

References

1. Dresselhaus, M. S.; Dresselhaus, G.; Saito, R. 1995. Physics of Carbon Nanotubes. *Carbon* 33, 883–891.
2. Wong, H.-S. P.; Akinwande, D. 2011. *Carbon nanotube and graphene device physics*. Cambridge University Press, Cambridge, UK.
3. Rafii-Tabar, H. 2008. *Computational physics of carbon nanotubes*. Cambridge University Press, Cambridge, UK.
4. Cummings, J.; Zetl, A. 2000. Low-friction nanoscale linear bearing realized from multiwall carbon nanotubes. *Science* 289, 602.
5. Rivera, J. L.; McCabe, C.; Cummings, P. T. 2003. Oscillatory Behaviour of Double-Walled Nanotubes under Extension: a Simple Nanoscale Damped Spring. *Nano Lett.* 3, 1001.
6. Rivera, J. L.; McCabe, C.; Cummings, P. T. 2005. Performance of Parallel TURBOMOLE for Density Functional Calculations. *Nanotechnology* 16, 186.
7. Zheng, Q.; Jiang, Q. 2002. Multiwalled Carbon Nanotubes as Gigahertz Oscillators. *Phys. Rev. Lett.* 88, 045503.
8. Levine, I. N. 2014. *Quantum chemistry*. Pearson, New York.
9. Mishra, B. K.; Ashok, B. 2018. Coaxial carbon nanotubes: from springs to ratchet wheels and nanobearings. *Mater. Res. Express* 5, 075023.
10. TURBOMOLE V6.2 2010, University of Karlsruhe and Forschungszentrum, Karlsruhe GmbH, Germany.
11. V. Arnim, M.; Ahlrichs, R. 1998. Performance of Parallel TURBOMOLE for Density Functional Calculations. *J. Comp. Chem.* 19, 1746.
12. Grimme, S.; Antony, J.; Ehrlich, S.; Krieg, H. 2010. A Consistent and Accurate ab initio Parametrization of Density Functional Dispersion Correction (DFT-D) for the 94 Elements H-Pu. *J. Chem. Phys.* 132, 154104.

13. Lebedeva, I. V.; Lebedev, A. V.; Popov, A. M.; Knizhnik, A. A. 2017. Comparison of Performance of van der Waals-corrected Exchange Correlation Functionals for Interlayer Interaction in Graphene and Hexagonal Boron Nitride. *Comput. Mater. Sci.* 128, 45.
14. Eichkorn, K.; Treutler, O.; Ohm, H.; Haser, M.; Ahlrichs, R. 1995. Auxiliary basis sets to approximate Coulomb potentials. *Chem. Phys. Lett.* 240, 283.
15. Sierka, M.; Hogekamp, A.; Ahlrichs, R. 2003. Fast evaluation of the Coulomb potential for electron densities using multipole accelerated resolution of identity approximation. *J.Chem. Phys.* 118, 9136.
16. Levshov, D.; Than, T. X.; Arenal, R.; Popov, V. N.; Parret, R.; Paillet, M.; Jourdain, V.; Zahab, A. A.; Michel, T.; Yuzyuk, Y. I.; Sauvajo, J.-L. 2011. Experimental evidence of a mechanical coupling between layers in an individual double-walled carbon nanotube. *Nano Lett.* 11, 4800.
17. Ghedjatti, A.; Magnin, Y.; Fossard, F.; Wang, G.; Amara, H.; Flahaut, E.; Lauret, J.-S.; Loiseau, A. 2017. Structural properties of double-walled carbon nanotubes driven by mechanical interlayer coupling. *ACS Nano* 11, 4840.
18. Cox, B. J.; Thamwattana, N.; Hill, J. M. 2008. Mechanics of nanotubes oscillating in carbon nanotube bundles. *Proc. R. Soc. A* 464, 691.
19. Zavalniuk, V.; Marchenko, S. 2011. Theoretical analysis of telescopic oscillations in multi-walled carbon nanotubes. *Low Temp. Phys.* 37, 336.
20. Fennimore, A. M.; Yuzvinsky, T. D.; Han, W.-Q.; Fuhrer, M. S.; Cumings, J.; Zettl, A. 2003. Rotational actuators based on carbon nanotubes. *Nature* 424, 408.
21. Hirschmann, T. C.; Araujo, P. T.; Muramatsu, H.; Zhang, X.; Nielsch, C.; Kim, Y. A.; Dresselhaus, M. S. 2013. Characterization of bundled and individual triple-walled carbon nanotubes by resonant. *ACS Nano* 7, 2381.
22. Kim, Y. A.; Muramatsu, H.; Hayashi, T.; Endo, M.; Terrones, M.; Dresselhaus, M. S. 2006. Fabrication of High-Purity, Double Walled Carbon Nanotube Buckypaper. *Chem. Vap. Deposition* 12, 327–330.
23. Bourlon, B.; Christian Glattli, D.; Miko, C.; Forr, L.; Bachtold, A. 2004. Carbon Nanotube Based Bearing for Rotational Motions. *Nano Lett.* 4, 709.
24. Kis, A.; Jensen, K.; Aloni, S.; Mickelson, W.; Zettl, A. 2006. Interlayer Forces and Ultralow Sliding Friction in Multiwalled Carbon Nanotubes. *Phys. Rev. Lett.* 97, 025501.
25. Silvestre, N. (ed.). 2016. *Advanced computational nanomechanics*. Wiley, Chichester.

7

Meshes and Networks of Multiwalled Carbon Nanotubes

K.R.V. Subramanian

Department of Mechanical Engineering
Ramaiah Institute of Technology
Bangalore, India

B.V. Raghu Vamshi Krishna and A. Deepak

Department of Mechanical Engineering
GITAM University
Bangalore, India

Aravinda C.L. Rao

Product Application & Research Center
Reliance Industries Ltd.
Vadodara, India

T. Nageswara Rao

Department of Mechanical Engineering
GITAM University
Bangalore, India

Raji George

Department of Mechanical Engineering
Ramaiah Institute of Technology
Bangalore, India

7.1 Introduction

This chapter deals with the subject field of multiwalled carbon nanotubes (MWCNTs) and their use in high-speed electrical and electronic circuits. Flexible electronics have been regarded as the next generation in optoelectronic devices in various areas, such as 3D molded interconnect devices, flexible displays, wearable electronics, organic light-emitting diodes, organic solar cells, and radio-frequency identification chips and circuits.

In recent years, conducting polymers [1], graphene films [2–4], CNT networks [5–7], conductive nanowires [8–11], and metal meshes have been widely studied as alternative materials for transparent flexible electrodes in both industry and academia. Among them, patterned silver mesh is a promising candidate due to its good transparency, conductivity, and flexibility. Currently, thermal R2R imprint lithography, roll offset printing process, and transfer printing process are existing technologies to fabricate the transparent metal mesh electrodes.

Lee et al. [12] demonstrated a high-durable AgNi nanomesh that exhibited strong adhesion by using simple transfer printing. Jang et al. [13] fabricated a metal (Ag) grid/AgNW hybrid transparent conducting electrode film composed of an electroplated metal grid and surface-embedded AgNW networks with excellent optoelectrical performance (transparency of 87% and sheet resistance of 13 Ω/ cm^2). Khan et al. [14] proposed a new type of electroplated copper nanomesh with superior electrical conductivity and optical transmittance. Kwon et al. [15] used monolayers of polystyrene spheres with different diameters as the template to create Ag honeycomb mesh electrodes. Han et al. [16] invented a uniform self-forming metallic network as a high-performance transparent conductive electrode. However, it is still a big challenge to produce the large-area metal mesh with line widths of submicron or even nanoscale with continuous mode in nonvacuum environment.

Other applications of patterned MWCNTs are as follows: When a small electric field is applied parallel to the axis of nanotube, electrons are emitted at a very high rate from the ends of the tube. This is called field emission. This effect can easily be observed by applying a small voltage between two parallel metal electrodes, and spreading a composite paste of nanotubes on one electrode. A sufficient number of tubes will be perpendicular to the electrode so that electron emission can be observed. One application of this effect is the development of flat panel displays. Samsung in Korea is developing a flat-panel display using the electron emission of carbon nanotubes. Nanotubes serve as catalysts for some chemical reactions. For example, nested nanotubes with ruthenium metal bonded to the outside have been demonstrated to have a strong catalytic effect in the hydrogenation reaction of cinnamaldehyde. Due to their nanoscale dimensions, electron transport in carbon nanotubes will take place through quantum effects and will only propagate along the axis of the tube [17]. Because of this special transport property, carbon nanotubes are frequently referred to as "one-dimensional."

In this chapter, we have investigated the meshes or networks formed by MWCNTs naturally. Attempts are made to order or self-organize the random networks with the application of external stimuli. Possibilities are discussed with theoretical analysis. The novelty of this work is to use simple external triggers such as sonication and electric field and use them to form engineered networks of the nanotubes.

7.2 Experimental Details

MWCNTs were purchased from AdNano Ltd., Shimoga, India. Extensive images were taken by transmission electron microscopy (TEM) at Indian Institute of Science Micro and Nano Characterization facility, Bangalore, India. Raman spectroscopy was also done at the same facility.

For ordering the random networks, a small quantity of the nanotubes was dissolved in ethanol to make suspensions and sonicated for various times (from 1 to 15 min). Drops of suspensions were delivered on small conducting plates with two electrodes and instantaneous sheet resistance/ voltage measured. The drops were also delivered on TEM grids and images taken quickly well within the re-coagulation times. External stimuli for ordering was provided in the form of voltage ranging from 1 to 12 V applied onto the electrodes after the drops of suspensions were delivered to the conducting plates. These were subsequently measured for instantaneous sheet resistance/voltage and imaged in TEM.

For the sheet resistances/voltages measured, theoretical analysis is provided.

7.3 Results and Discussion

Figures 7.1–7.4 show some high-resolution TEM images of random meshes/networks of MWCNTs. It may be noted that electronic conductivity is good along the nanotube length and also at the junctions or meeting points. At the ends of the nanotubes that are open, dangling or unsaturated bonds contribute additional electrons for conductivity.

Figure 7.3 shows the number of random interconnections between the nanotubes. Also, Figure shows the essentially fractal nature of the nanotube network. The nanotubes may overlay each other or intersect at the junctions or provide multibranching from a common node or provide looped junctions. Such multifarious junctions are clearly delineated in Figures 7.5–7.7.

Raman spectra of the MWCNTs are shown in Figure 7.8. The G band is found at 1590 cm^{-1}, while the D band and 2D bands are found at 1350 and 2690 cm^{-1}, respectively. The I_d/I_g ratio is found to be 0.84. Radial breathing modes are observed at less than 230 cm^{-1}. Figure 7.9a, b shows the radial breathing Raman spectrographs.

A simple measurement setup (Figure 7.10) was made with a conducting bowl made of aluminum with two electrodes positioned inside the bowl at fixed separation (2 cm). The electrodes were the strips made of aluminum and there was a gap (0.1 cm) between the bottom of the electrode and the bowl base. The MWCNT nanosuspension in ethanol

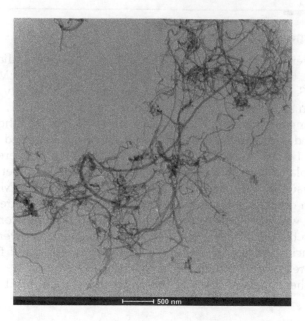

FIGURE 7.1
A high-resolution TEM image of a random mesh/network of MWCNTs, which is fractal in nature.

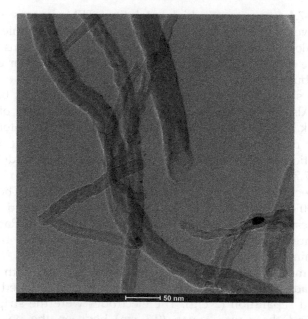

FIGURE 7.2
A high-resolution TEM close-up magnified image of a random mesh/network of MWCNTs, which is fractal in nature.

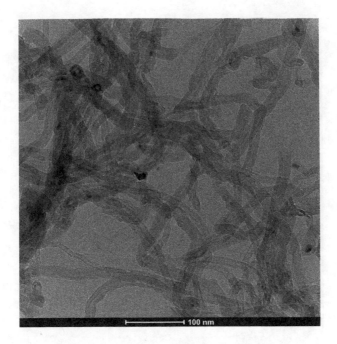

FIGURE 7.3
A high-resolution magnified TEM image of a random mesh/network of MWCNTs, which is fractal in nature. Note the number of random interconnections between the nanotubes.

(0.5 g of MWCNT in 2 ml of ethanol) was spread on the bowl so as to contact the two electrodes. External electrical connections were taken from the two electrodes. The nanosuspension was subjected to different pre-processing stimuli steps: (a) ultrasonication for varying lengths of time (1, 3, 5, 7, and 10 min), (b) electric field application (3.6 and 12 V), and (c) ultrasonication followed by electric field application.

After the pre-processing stimuli, the potential difference in mV was measured by connecting a digital multimeter across the electrodes. The potential difference was measured as the stable value after a 10 min duration allowing for the fluctuations to settle.

Capacitance is given by the formula:

$$C = Q/V \qquad (7.1)$$

As voltage or potential difference decreases, the capacitance increases.

Capacitance can also be represented as:

$$C = (\varepsilon_0 \times \varepsilon_r \times A)/d \qquad (7.2)$$

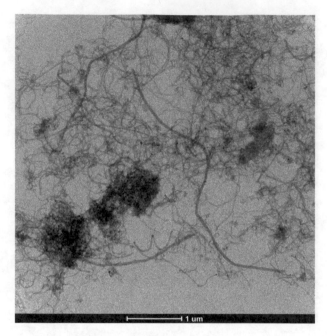

FIGURE 7.4
A high-resolution TEM image of a random mesh/network of MWCNTs, which is fractal in nature. Note the fractal nature of the network.

- As capacitance increases, the relative permittivity of the nanosuspension increases. The main purpose of ultrasonication and combined application of electric field is to increase the capacitance and relative permittivity.
- MWCNTs are piezoelectric. Ultrasonication triggers the piezoelectric property and electric field application causes the piezoelectric domains to align in the direction of field (a phenomenon called "poling"). The application of external field may cause existing domains to flip or change the direction resulting in negative voltage and resultant increase in the capacitance and relative permittivity of the nanosuspension.

Figures 7.11–7.14 show the effect of external stimuli on the resultant voltage (and capacitance) from the nanosuspensions. From Figure 7.11, it can be inferred that as time of ultrasonication increases, the mechanical agitation triggers the piezoelectric property and causes the piezoelectric domains to flip and reorient and give rise to a net negative potential difference. However, at long ultrasonication times (7 min and higher), the flipping effect cancels out and a random distribution of the

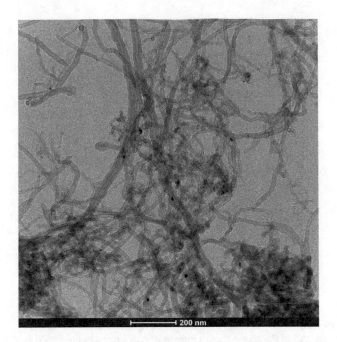

FIGURE 7.5
Another TEM image of the fractal network of MWCNTs.

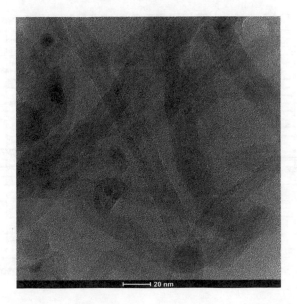

FIGURE 7.6
A high-resolution TEM close-up magnified image of a random mesh/network of MWCNTs, which is fractal in nature. Note that the nanotubes may overlay each other or intersect at junctions or provide multi-branching from a common node or provide looped junctions.

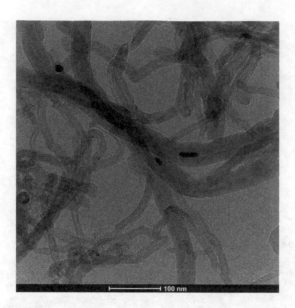

FIGURE 7.7
A high-resolution TEM close-up magnified image of a random mesh/network of MWCNTs, which is fractal in nature. All of the above type of junctions as described in Figure 7.6 as above are seen.

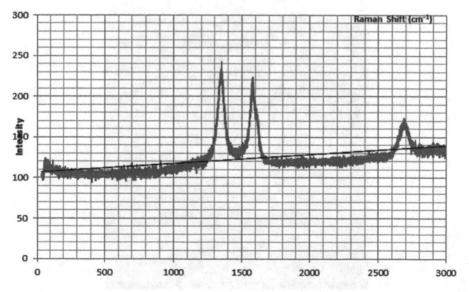

FIGURE 7.8
Raman spectroscopy data of MWCNTs.

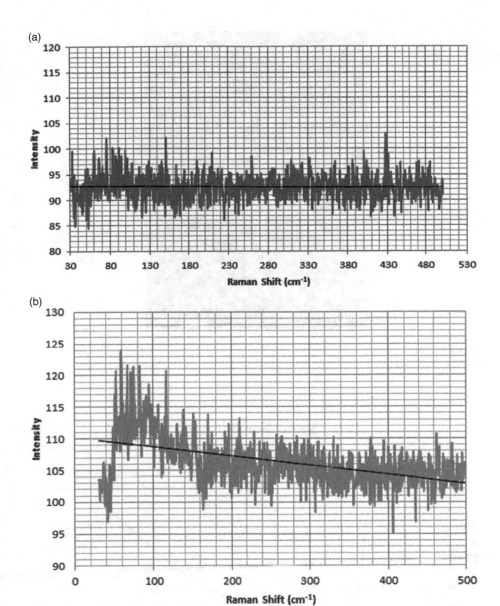

FIGURE 7.9
(a, b) Radial breathing modes of the MWCNTs.

FIGURE 7.10
Measurement setup.

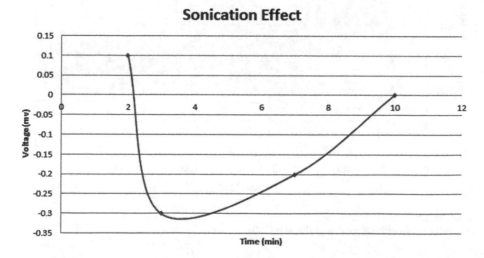

FIGURE 7.11
Effect of different ultrasonication times on the resultant voltage and potential difference from
the nanosuspension.

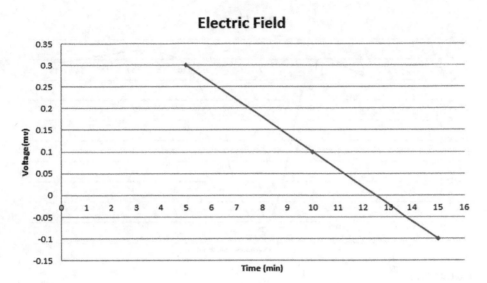

FIGURE 7.12
Effect of duration of application of electric field (3.6 V) on the resultant voltage and potential difference from the nanosuspension.

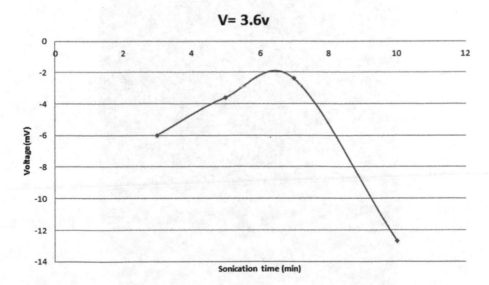

FIGURE 7.13
Effect of combined ultrasonication (for different times 3, 5, 7, and 10 min) and subsequent application of the electric field (3.6 V).

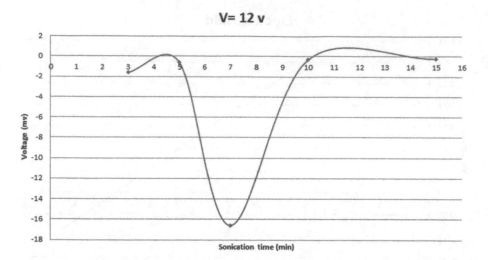

FIGURE 7.14
Effect of combined ultrasonication (for different times 3,5,7,10, and 15 min) and subsequent application of the electric field (12 V).

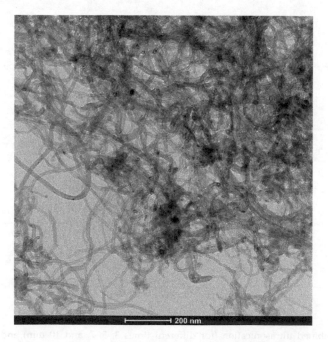

FIGURE 7.15
TEM image of sample where the nanosuspension was ultrasonicated for 7 min.

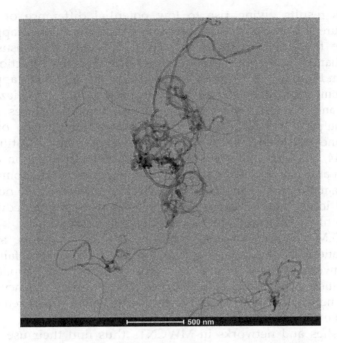

FIGURE 7.16
TEM image of sample where the nanosuspension was ultrasonicated for 10 min and then subjected to electric field (3.6 V) for 10 min.

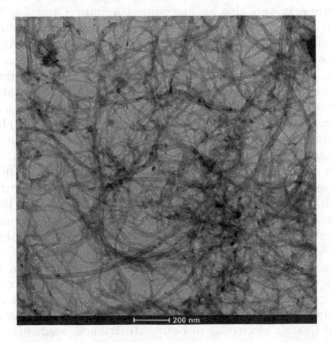

FIGURE 7.17
TEM image of sample where the nanosuspension was subjected to electric field (12 V) for 5 min.

nanotubes results giving rise to low potential difference or voltage. From Figure 7.12, it can be inferred that as the duration of application of electric field (3.6 V) increases, the CNTs in the nanosuspension acquire charges and show a tendency to orient in the direction of the field. From Figure 7.13, it can be inferred that the combined application of ultrasonication and electric field causes poling of piezoelectric domains and the resultant flipping of the domains causes negative voltage output. At short times of ultrasonication, extant of poling reduces and it increases only at long times of ultrasonication. From Figure 7.14, it can be inferred that the combined application of ultrasonication and electric field causes poling of piezoelectric domains and the resultant flipping of the domains causes negative voltage output. At short and long times of ultrasonication, extant of poling reduces and increases only at intermediately long times of ultrasonication.

From TEM images shown in Figures 7.15–7.17, it can be seen that ultrasonication or electric field application alone produces a fairly interwoven network of the nanotubes. However, the combined application of ultrasonication and electric field (Figure 7.16) shows a tendency to fragment the network and produce smaller fragments with a tendency to orient in the direction of the field.

The meshes and networks in MWCNTs thus find their use as alternatives to silver and copper meshes for high-speed circuits. MWCNTs naturally form random networks and these are shown in this chapter. It is possible to order the meshes/networks because of their self-aligning property and also by mechanical agitation and application of an external electric field. These possibilities are discussed with theoretical analysis. Lee et al. [12] demonstrated a high-durable AgNi nanomesh that exhibited strong adhesion by using simple transfer printing. Jang et al. Fabricated a metal (Ag) grid/AgNW hybrid transparent conducting electrode film composed of an electroplated metal grid and surface-embedded AgNW networks with excellent optoelectrical performance (transparency of 87% and sheet resistance of 13 Ω/cm^2) have been demonstrated. Khan et al. proposed a new type of electroplated copper nanomesh with superior electrical conductivity and optical transmittance. Kwon et al. used monolayers of polystyrene spheres with different diameters as the template to create Ag honeycomb mesh electrodes. It is still a big challenge to produce the large-area metal mesh with line widths of submicron or even nanoscale with continuous mode in nonvacuum environment. In this study, we have investigated the meshes or networks formed by MWCNTs naturally.

Attempts are made to order or self-organize the random networks with the application of external stimuli. In recent years, conducting polymers, graphene films, CNT networks, conductive nanowires, and metal meshes have been widely studied as alternative materials for transparent flexible electrodes in both industry and academia.

7.4 Conclusion

A fractal network or mesh of MWCNTs is analyzed. MWCNTs naturally form random networks and these are shown in this chapter. It is possible to order the meshes/networks because of the self-aligning property and also by mechanical agitation and application of an external electric field. These possibilities are discussed with theoretical analysis.

References

1. Wu H et al., "A transparent electrode based on a metal nanotrough network," *Nat. Nanotechnol.*, **2013**, vol. 8, no. 6, pp. 421–425.
2. Bae S et al., "Roll-to-roll production of 30-inch graphene films for transparent electrodes," *Nat. Nanotechnol.*, **2010**, vol. 5, no. 8, pp. 574–578.
3. Wang J, Liang M, Fang Y, Qiu T, Zhang J, and Zhi L, "Rod-coating: towards large-area fabrication of uniform reduced graphene oxide films for flexible touch screens," *Adv. Mater.*, **2012**, vol. 24, no. 21, pp. 2874–2878.
4. Li X et al., "Large-area synthesis of high-quality and uniform graphene films on copper foils," *Science*, **2012**, vol. 324, no. 5932, pp. 1312–1314.
5. Tenent RC et al., "Ultrasmooth, large-area, high-uniformity, conductive transparent single-walled-carbon-nanotube films for photovoltaics produced by ultrasonic spraying," *Adv. Mater.*, **2009**, vol. 21, no. 31, pp. 3210–3216.
6. Wu Z et al., "Transparent, conductive carbon nanotube films," *Science*, **2004**, vol. 305, no. 5688, pp. 1273–1276.
7. Liu Y et al., "Enhanced photoelectrochemical properties of Cu2 O-loaded short TiO2 nanotube array electrode prepared by sonoelectrochemical deposition," *Nano-Micro Lett.*, **2010**, vol. 2, no. 4, pp. 277–284.
8. van de Groep J, Spinelli P, and Polman A, "Transparent conducting silver nanowire networks," *Nano Lett.*, **2012**, vol. 12, no. 6, pp. 3138–3144.
9. Hu L, Kim HS, Lee JY, Peumans P, and Cui Y, "Scalable coating and properties of transparent, flexible, silver nanowire electrodes," *ACS Nano.*, **2010**, vol. 4, no. 5, pp. 2955–2963.
10. Kiruthika S, Gupta R, Rao KDM, Chakraborty S, Padmavathy N, and Kulkarni GU, "Large area solution processed transparent conducting electrode based on highly interconnected Cu wire network," *J. Mater. Chem. C*, **2014**, vol. 2, no. 11, pp. 2089–2094.
11. Lee P et al., "Highly stretchable and highly conductive metal electrode by very long metal nanowire percolation network," *Adv. Mater.*, **2012**, vol. 24, no. 25, pp. 3326–3332.
12. Kim HJ et al., "High-durable AgNinanomesh film for a transparent conducting electrode," *Small*, **2014**, vol. 10, no. 18, pp. 3767–3774.
13. Jang J, Im HG, Jin J, Lee J, Lee JY, and Bae BS, "A flexible and robust transparent conducting electrode platform using an electroplated silver grid/surface-embedded silver nanowire hybrid structure," *ACS Appl. Mater. Interfaces*, **2016**, vol. 8, no. 40, pp. 27035–27043.

14. Khan A et al., "High-performance flexible transparent electrode with an embedded metal mesh fabricated by cost-effective solution process," *Small*, **2016**, vol. 12, no. 22, pp. 3021–3030.
15. Namyong K, Kyohyeok K, Sihyun S, Insook Y, and Ilsub C, "Highly conductive and transparent Ag honeycomb mesh fabricated using a monolayer of polystyrene spheres," *Nanotechnology*, **2013**, vol. 24, no. 23, Art. no. 235205.
16. Han B et al., "Uniform self-forming metallic network as a highperformance transparent conductive electrode," *Adv. Mater.*, **2014**, vol. 26, no. 6, pp. 873–877.
17. Hong, S, and Myung S., "Nanotube electronics: a flexible approach to mobility," *Nat. Nanotechnol.*, 2007, vol. 2, pp. 207–208. doi:10.1038/nnano.2007.89.

Index